S0-BKE-715

ANIMALS
OF THE WORLD
NORTH AMERICA

Eric Powell
Bertel Bruun
Devra Kleiman
Brendan Kypta

ANIMALS OF THE WORLD NORTH AMERICA

HAMLYN

Published by
THE HAMLYN
PUBLISHING GROUP LTD
LONDON · NEW YORK · SYDNEY · TORONTO
Hamlyn House, Feltham
Middlesex, England

SBN 600 10016 2

Photoset by BAS Printers Ltd
Printed in Hong Kong

CONTENTS

Fishes

Eric Powell is a professional writer on ichthyology whose experience as a practical aquarist and angler enable him to write with real authority on the subject. He has travelled to several countries to learn at first hand about their fish fauna. His interest in the fishes of North America has been developed through both study and practical experience.

Birds

Bertel Bruun is a Danish-born medical doctor who has taken up residence in New York. His knowledge of birds is well known; he has written a handbook on the birds of Europe, contributed to a field guide to the birds of North America, and recently completed work on a major project dealing with British and European birds. He has come to be known as an authority on birds and bird behaviour in both Europe and North America.

Mammals

Devra Kleiman is a young and enthusiastic American graduate who has spent some years working in a London research institute. Her special interests are centred around the small mammals and in particular animal behaviour. The knowledge she has gained during recent years has given her a first-hand insight into the fascinating behaviour patterns of mammals and the effect they have on the animal communities around them.

Reptiles

Brendan Kypta is the pen-name of a professional zoologist who has been deeply interested in reptiles since an early age, and his career has taken him to many parts of the world to catch and study these animals. Now retired from full-time work as a herpetologist, he is writing up his vast experience in books and magazine articles as well as making occasional television and radio broadcasts.

NORTH AMERICA

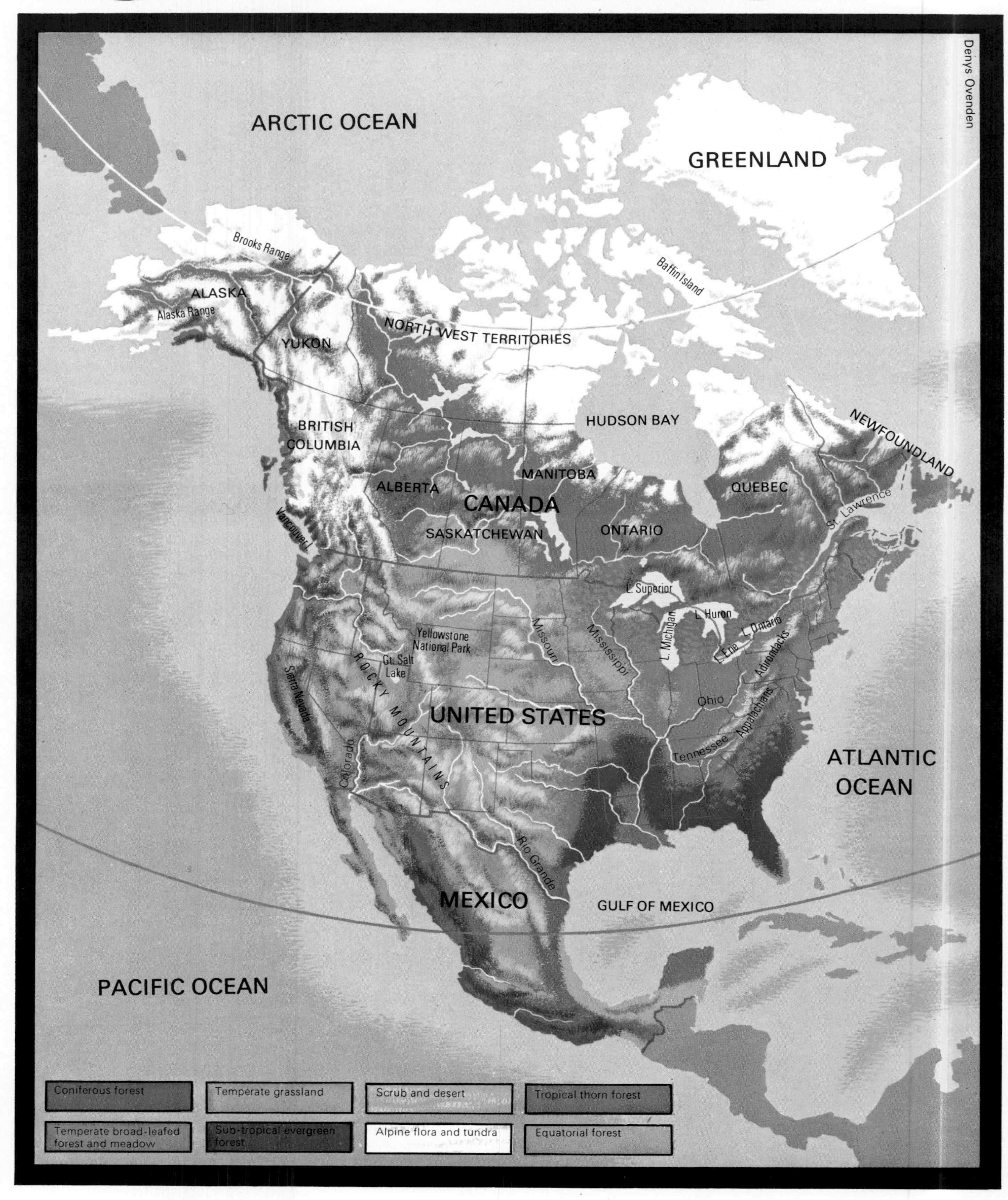

The Americas and their early discovery is a subject that although well investigated will probably never be explained to the satisfaction of everyone.

The continent as a whole is so large (over sixteen million square miles) that it is unlikely that anyone will be in a position to say that man first entered the continent at this or that point. It would certainly seem that the first 'settlers' came from the east and when they arrived they undoubtedly found an abundance of fauna and flora. The enormous size alone of North America means that the temperature and other climatic conditions vary tremendously from place to place. There are vast areas that are covered in ice and snow the whole year round, and desert areas where rain never falls on the dry parched ground and barren rocks, yet both these areas have some wildlife.

Between these greatly differing climatic conditions lie gradations of conditions that give rise and protection to large numbers of animals and plants. Many of the areas can be described as lush and tropical in appearance although no part of the continent is strictly tropical. Everyone is aware of the fantastic development that has been achieved in North America in order to benefit the human inhabitants. Huge cities, enormous highways and agricultural development, besides all the mineral workings, have opened up the land to the benefit of man. Not everyone is aware of the amount of thought and work that has gone into the protection of wildlife which of course is also for the benefit of man. The various bodies within Canada and the United States that are striving to protect the natural fauna and flora have for almost 100 years worked in collaboration with each other in order to achieve maximum protection.

The numbers of people supporting wildlife preservation are increasing rapidly, children at school are soon made to realize that where their homes now stand wild animals once roamed in perfect freedom and that these animals are now rare if not altogether extinct.

It is to the immense credit of the founders of Yellowstone Park that they formed the first national wildlife park in the world, and that since then other nations have taken up these ideals and are constantly examining new areas for wildlife parks, making it possible for future generations to study their own inherited wildlife.

As with other books in this series, the authors are usually natives of North America who are making a special study of the animals they are writing about; in other cases the authors have spent long periods of research in North America and have returned to their homes to write up their findings.

Almost every author would wish to have more space for his or her text and photographs and, as in this book, there is never sufficient room to include everything. Some beautiful and very interesting animals such as amphibians and insects have had to be omitted, but with chapters on fishes and reptiles as well as mammals and birds, the book provides an excellent introduction to the animals of this important continent.

FISHES

Eric Powell

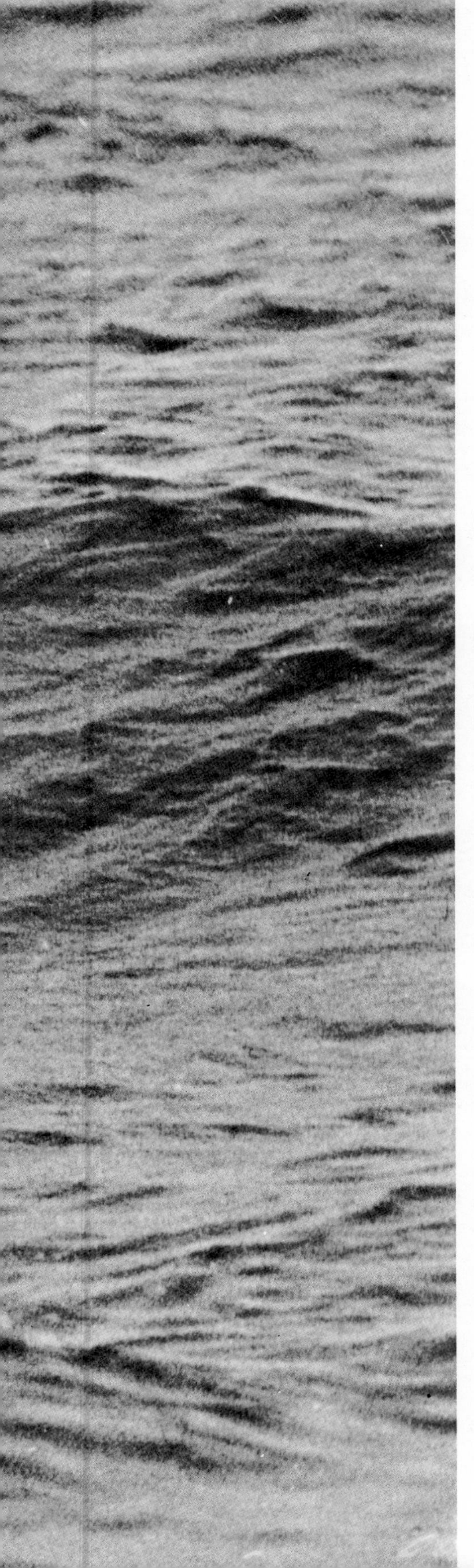

The early European settlers in North America faced many problems in their new world, and among them was the general unfamiliarity of the wildlife around them. They knew the names of very few of the fish they caught, and only by trial did they find out whether they were good to eat, tasteless or even poisonous (and the trial of the poisonous kinds could be very painful and even fatal). Here and there, peaceful contact with the aboriginal inhabitants, who in general depended very heavily on the 'sea food' of the coasts and the fishes inland, taught the immigrants their names for local fish. Very many more fish were named because of their similarity to those that were familiar in Europe, even if they were only remotely, if at all, related to their Old World counterparts. As a result there are today on the Atlantic coast a number of 'flounders' which although flatfishes, are only distantly related to the European flounder, and 'hakes' which are not hakes at all to European eyes. Similarly, inland, there are a whole host of 'minnows' and 'chubs', named for their general resemblance to the fishes of England known by these names.

Settlers in North America did not, however, find every fish strange and unknown. Some were old friends from the seas and inland waters of Europe. The fishes must have presented a considerable paradox in that the 'New' World contained so many kinds that were well known in the 'Old' World.

One of the most primitive groups of fishes known are the cyclostomes. Belonging to the class Agnatha (the jawless animals) they can be said not to be fishes at all; but because they live in water, are eel-like in shape, and can be eaten, we usually treat them as fishes. The most familiar of the North American forms is the Sea Lamprey (*Petromyzon marinus*), a three-foot, yellow mottled, slimy, eel-like creature which has made itself only too well known in the Great Lakes on the borders of Canada and the northern U.S.A. Its funnel-like mouth is covered with rows of blunt, conical yellowish teeth, and lacking jaw bones, it forms a most effective sucker disc by which it attaches itself to a living fish. A muscular tongue, also armed with teeth, rasps away the scales and skin of the victim, while the lamprey's salivary glands produce a secretion which, pumped into the wound, prevents the fish's blood from clotting. The lamprey sucks blood from its victim with tremendous power until sated, when it releases its hold while the attacked fish is left to die from its anaemia or injuries, and very rarely to recover.

The Atlantic Sailfish (Istiophorus albicans) *may reach 11 feet long and weigh 120 lbs. On the west coast of America it is replaced by the Pacific Sailfish* (Istiophorus orientalis).

Lampreys have a series of seven small gill holes on the sides of the head, and a single nostril between their eyes. Although found in the sea relatively commonly, they breed in fresh water. On the east coast they have become rather uncommon due to increasing pollution and damming of many of the rivers. The lamprey is also found landlocked in some lakes, or in other words it treats the lake as if it were the sea and enters tributary rivers to spawn. The lamprey was found originally in Lake Ontario but never penetrated to the upstream lakes because of the impassable Niagara Falls. Then in 1828, the Welland canal was cut joining Lakes Ontario and Erie and affording a passage upstream. Nothing happened until 1932 when lampreys were found in Lake Erie; five years later they were reported in Lakes Huron and Michigan, and by 1946 they had reached Lake Superior. The effects were disastrous. Offered a virgin lake with no predators and abundant fish, the lampreys swarmed throughout the lakes. Valuable fisheries were ruined, game fishing declined and many species became rare as a result of the abundance of the lampreys.

Among the fishes of the near Arctic regions of North America are a number of other fish which can be found also in Europe. One of these is the Pike (*Esox lucius*) that large predatory

Above: remoras (family Echeneidae); one is attached and another about to attach itself to a Hawksbill Turtle (Chelone imbricata).
Right: Northern Pike (Esox lucius).
Below: the Texas Cichlid (Herichthys cyanoguttata), *the only member of its family to reach as far north as the U.S., is a colourful aquarium fish rarely kept in captivity due to its voracious habits.*
Far right top: Smallmouth Black Bass (Micropterus salmoides) *showing the single spiny dorsal fin typical of the family.*
Far right bottom: the Rainbow Trout (Salmo gairdneri).

fish of lakes and rivers, with flattened snout and innumerable teeth. This same species is found throughout northern America, north of the parallel of Ohio State, except for the extreme west. In America it is known as the Northern Pike, to distinguish it from other members of the family Esocidae, for there are four in all in these regions. Two of the American species, both relative midgets, are known as pickerels; the Chain Pickerel (*Esox niger*) which grows to a length of twenty-four inches, and the Grass Pickerel (*Esox americanus*) at twelve inches are both found in the eastern States.

The pike of the United States, in the eyes of the game fishermen, is however the Muskellunge, or 'Muskie' (*Esox masquinongy*). Its maximum weight has been reported to be 102 pounds, but it is relatively restricted in distribution being found only around the Great Lakes region.

At first glance there would not seem to be much in common between the giant, predatory Muskellunge, and the abundant, three-inch-long American mud-minnows, but the two are closely related, and both belong to the order Haplomi. Mud-minnows are heavily scaled, hump-backed little fish with blunt snouts, which live along the sides of densely vegetated streams, lakes and ponds. Several species are known, but all have a remarkable tolerance to oxygen-deficient conditions, and they can burrow and survive in the stickiest of mud bottoms during droughts. The Eastern Mud-minnow (*Umbra pygmaea*) which is found along the coastal fringe from New York to Florida, is the smallest of the three species known from America.

Another relative, the Alaska Blackfish (*Dallia pectoralis*) looks very similar to the mud-minnows, and indeed takes their place in the cold regions of Arctic America. It grows to around eight inches in length, and as it lives in rivers which are frozen for a large part of the year it has an amazing resistance to cold. Many are the tales of blackfish found frozen in ice swimming actively when thawed out, although few of them record how long the fish survived.

The Salmon (*Salmo salar*) is another fish familiar to European eyes which is found along the east coast of North America from the coast of Labrador southwards to the Connecticut River. When Europeans first came to New England and the Maritime Provinces they found Salmon in every waterway which was not obstructed by impassable falls. In New England alone there were at least nineteen sizeable Salmon rivers, but dams were built and the waters used industrially, so that by the 1880s only eight had Salmon runs. This failed in one of the eight by 1895, and by 1925 only two rivers in Maine had regular Salmon runs. Salmon are, of course, particularly susceptible to destruction in this way, as they breed in fresh water usually well upstream, but spend a sizeable part of their life in the sea. Any particularly heavy pollution in the estuary or obstruction in the shape of badly designed dams will, over a period of years, extinguish the population in a river.

The well known Rainbow Trout (*Salmo gairdneri*) which has been introduced widely into the British Isles and Europe as well as America outside its original range, is a trout which was found in western America from southeastern Alaska to California. It is now found in many places in the United States and southern Canada, and its game fighting qualities, good growth rate, and above all its edibility, have made it the ideal fish to introduce to angling waters. The Rainbow Trout is found as two forms. A wholly freshwater form (the typical Rainbow Trout, formerly known as *Salmo irideus*) lives in lakes and streams and spawns in their tributaries in early spring, and only rarely grows heavier than five pounds. The second form, which likewise spawns in early spring, migrates to the sea after spawning, while the young remain in fresh water for two or three years. This migratory form grows very much larger and may weigh up to thirty pounds, although eight to nine is a good average. It too has the beautiful pinkish 'rainbow' along its side, but is rather more silvery than the fresh-water form. Because of its colour, and its migratory

Right: the Cod (Gadus callarias) *is widely distributed in the North Atlantic, westwards to Newfoundland and the U.S. Far right: the Alaska Blackfish* (Dallia pectoralis) *is abundant in far northern areas, where it is able to withstand the intense cold of the arctic winter.*

habits, it was thought at one time to be a distinct species, *Salmo gairdneri*, and sea run fish are still known as Steelhead Trout.

Another western American trout found from northern California through western Canada to Alaska is the Cut-throat Trout (*Salmo clarki*), so called because of a red streak under the lower jaw on the throat. In places the Cut-throat is non-migratory, but mostly they descend to the sea as young fish and do not return until adult, when in autumn they enter the rivers to spawn in the spring. The Cut-throat is also a good game fish growing to thirty inches and a weight of seventeen pounds.

The Pacific coast of North America is also the home of the several Pacific salmons (*Oncorhynchus spp.*) which form the basis of the very important salmon fisheries of the U.S.A. and Canada. Five American species are known, which have a variety of confusing common names, and apparently unpronounceable scientific names. All, however, have several general features in common. They all spawn in the rivers or lakes in winter, all migrate downstream and spend two or more years in the sea on the rich feeding grounds of the North Pacific, feeding on sand-eels, herrings, squids and pelagic crustaceans. All are found in the North Pacific basin, roughly from northern California to the Arctic and on the western side south to China and Japan. There are variations between the species in their biology. Some like the Sockeye (*Oncorhynchus nerka*), spawn hundreds of miles upstream from the coast, entering the rivers in spring or summer and spawning the following winter, while others such as Humpback or Pink Salmon (*Oncorhynchus gorbuscha*), and the Chum (*Oncorhynchus keta*) spawn only just upstream of the estuary, having entered the river only a month or two before spawning. These and the two other North American salmons, the Chinook (*Oncorhynchus tshawytscha*) and the Coho (*Oncorhynchus kisutch*), are commercially fished, both at sea using floating nets near the surface, and by netting and trapping in and near the estuaries as they enter the rivers. Much of the catch is canned, although in recent years freezing has been widely used.

Despite the very considerable fishing industries along the American Pacific coast there still seems to be an immense abundance of fish of various kinds, and compared to other oceans its resources are largely untapped. Notable among these commercially exploited fishes is the Pacific Herring (*Clupea harengus pallasi*), a close relative and usually regarded as a sub-species only of the Atlantic Herring (*Clupea harengus*). This is widespread in the North Pacific, spawning in spring in four major centres on the Canadian coast, mostly close inshore, often in the estuaries and low salinity areas of this region.

Further south a relative of the herring takes its place as the subject of a once important fishery. This is the Californian Sardine (*Sardinops sagax*), a different fish altogether from the European Sardine or Pilchard (*Sardina pilchardus*) although closely related to that found off the South African coast. The history of the Californian Sardine industry is not old. It started during the First World War in response to increased demand for canned sardines, and the prevention of fishing in Europe. The total catch in 1915 was four million pounds, by 1918 it was 158 million pounds, by 1929 landings totalled 652 million pounds and in 1936 a scarcely credible 1,500 million pounds were landed. From then until 1944 a sharp decline took place and by 1953 the sardine fishery was almost

defunct. The vast quantities of sardines taken from the waters round California (the number of fish caught in 1939 was estimated at over 7,000 million) had decreased the population to below a viable level, and as a result changes in migrations and abundance took place which meant that the fish were mostly further offshore. Today, the sardine is still present on the Pacific coast from Baja California to British Columbia, but the fishery is only a fraction of what it was in its heyday.

The introduction of the American Shad (*Alosa sapidissima*) from its natural haunts in the northern Atlantic to the Pacific coast is a story with a different ending. This shad, a large relative of the herring, growing regularly up to eight pounds in weight, was abundant on the Atlantic coast, spending most of its life in the sea but entering fresh water to spawn and returning to the sea after spawning. They are a shoaling fish rarely found deeper than fifty fathoms, and feeding mainly on plankton. Like the Salmon, they are very vulnerable to pollution and obstructions in rivers, but they were present in teeming millions, and still are on a reduced scale, and formed a valuable fishery along the coast from Newfoundland to Florida. In 1871 eggs were first carried across to the west coast, and the shad successfully introduced to the Pacific. They never looked back. Today, the American Shad is found from southern California to southern Alaska and it still appears to be spreading. It is now found abundantly enough to be fished for, and is a fine flavoured food fish, even if rather bony.

Another fish which was introduced, this time to America from Europe, and one which did not have such happy consequences, is the Carp (*Cyprinus carpio*). In 1870 Carp were introduced into the United States. No doubt they were released there for what seemed the best possible reasons, one that they would be a valuable food fish in inland America. From the naturalist's viewpoint, and as it turned out the fisherman's interests, it was a disaster. The Carp is today one of the most important fresh-water fishes outside the cold regions of northern America, not for its food value, market price or sports appeal, but from the terrible impact it has had on the fresh waters in which it lives. Carp live in the shallows and feed very largely on plants. They have found the climate and habitats of America very suitable, and browse down the vegetation which sheltered so many of the native species. It has even proved a more than successful competitor with American trout in the cooler areas. Not only has it proved a threat to many of the native fish of restricted distribution, but it has even affected the wildfowl populations on the marshes by eating the ducks' food plants.

Native relatives of the Carp are very abundant, however, without the introduction of Old World species. Many of the 190-odd species are known as 'minnows' or 'chubs' and there are a very large number of 'shiners', all of which are endemic American fish. Some, in fact are extremely limited in their distribution like the twelve-inch Splittail (*Pogonichthys macrolepidotus*) and the Hardhead (*Mylopharodon conocephalus*) which may grow to three feet, both of which are found only in the Sacramento River of California.

The largest 'minnow' in North America is the Colorado Squawfish (*Ptychocheilus lucius*) which is native only to the Colorado River and its tributaries from Mexico to Wyoming. In former times it was very much more abundant than it is today and it was a valuable source of food for the Indians and the early settlers in the area. Reports have it that it grew to a

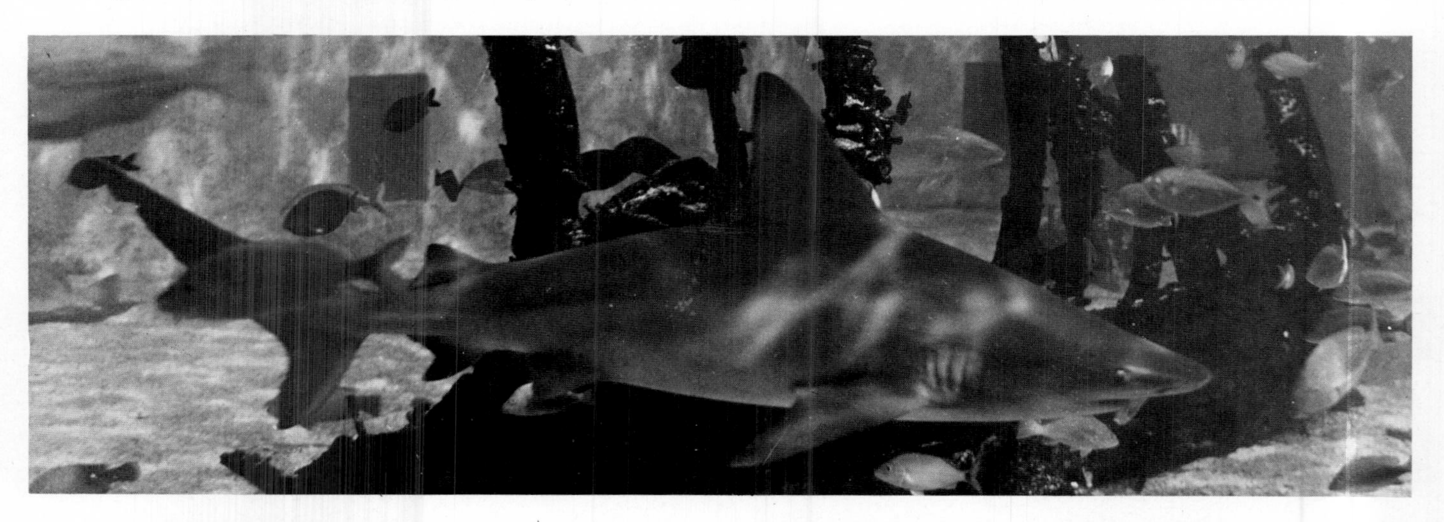

Above: Sandbar Shark (Carcharhinus milberti) *showing its superbly streamlined body, tiny eyes and small gill slits.*
Top: the Atlantic Tarpon (Megalops atlanticus) *is famed as a sporting fish; it ranges as far north as Cape Cod and sometimes enters rivers.*
Centre right: top view of remora (family Echeneidae) showing the sucker disc by which it attaches itself to the underside of sharks, turtles and sometimes skin divers. Remoras are not parasites, but glean scraps of food from their hosts.
Far right: Bluntnose Stingray (Dasyatis sayi). *The poison spine on its tail can inflict a dangerous wound on a diver.*

eastern Mexico, and from the Rocky Mountains in the west to the Appalachians in the east. It grows to a maximum length of three inches and is abundant in a wide variety of habitats from sluggish brooks, lakes and ponds, to cool and warm water. The eggs are deposited on the underside of a flat object or hollow rock in a carefully chosen site. Several females may spawn in the nest and the male guards the eggs, swimming constantly backwards and forwards and lightly brushing the eggs with a spongy pad on its back to free them from settled silt. This careful protection of the eggs is probably partly responsible for the fish's success and widespread distribution. It is a favourite food for many larger food and game fish, and it is propagated widely in fish hatcheries as a food fish for the larger species reared in the hatchery. It is also used as a bait by anglers.

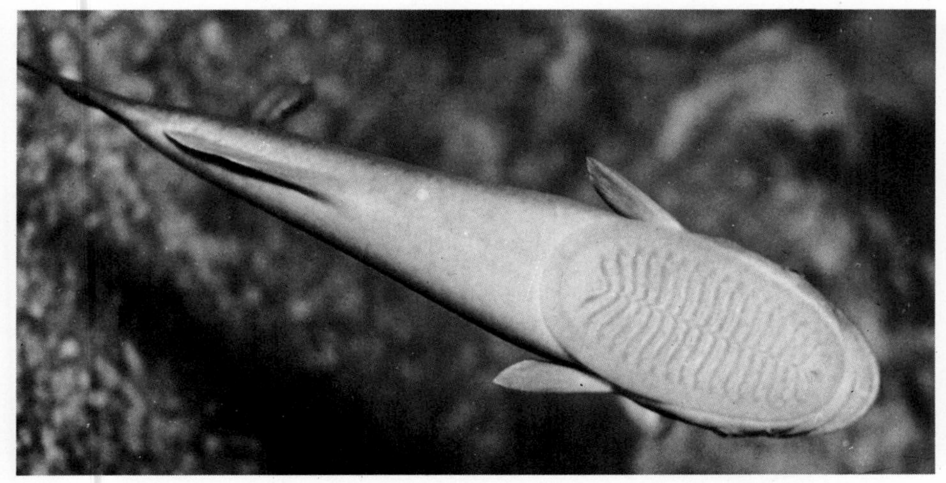

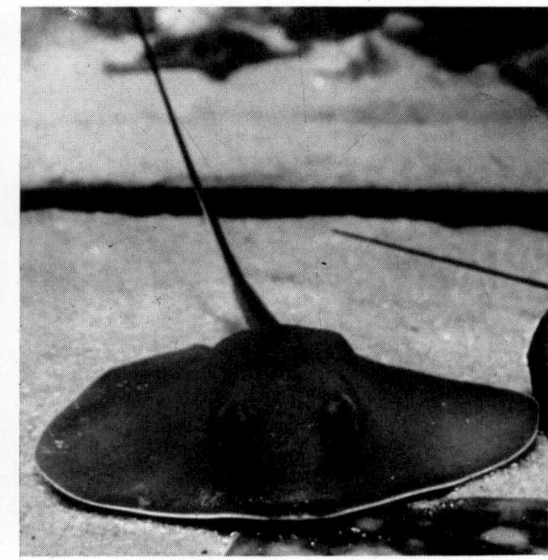

length of about six feet, and a weight of 100 pounds, but the more recent captures of the fish were only up to half this weight. The dams constructed in the rivers appear to have prevented the squawfish from making its major spawning runs upstream and those which still survive are mere isolated populations. The squawfish lives in the big rivers where the current is strong, the water muddy, the bottom stony or muddy and about three feet deep. The Indians used to capture them with a primitive dip net in shallows, or as they were stranded by receding floods. Later they were shot with bow and arrow, or speared. Certainly this is in agreement with their habit of lying in deep holes, particularly below rapids, and feeding on debris carried down the river. They are carnivorous, and can be caught on baits as large as a young rabbit. Related squawfishes are found elsewhere in North America.

A more minnow-like 'minnow' is the Fathead Minnow (*Pimephales promelas*) which is widely distributed in the area from southern Canada to north-

Another group of relatives of the carp fishes, which are peculiarly North American, are the suckers (family Catostomidae). The suckers get their name from their protrusible mouth and fleshy lips on the underside of the head. Some sixty-five species are known, some of which grow to a length of three feet, although the majority are much smaller. Among the larger species are the Bigmouth Buffalo (*Ictiobus cyprinellus*) which grows up to three feet long and is one of the few suckers in which the mouth is set at the

end of the head, and not on the underside. It is typically an inhabitant of large rivers, oxbows and shallow lakes, where it feeds on molluscs, insect larvae and vegetation, and is found from Saskatchewan to the Ohio Valley, and south to Alabama, Louisiana and Texas. It is fished for commercially, along with the Quillback (*Carpiodes cyprinus*) which grows to around the same size. The Quillback is found in much the same area as the Bigmouth Buffalo, living in large rivers and lakes near the shore where the bottom is silty or sandy. It is a very distinctive sucker with the first rays of the dorsal fin long and flagpole-like.

One of the most distinctive of the suckers in North America is the Humpback Sucker (*Xyrauchen texanus*) which is found in the lower parts of the Colorado River system. It occurs commonly in four to eight feet of water over a bottom of sand or rock. Where it lives the current is usually appreciable and submerged water plants do not grow; consequently it has to live in the force of the current with very little shelter. The distinctive and sharp-edged hump which forms a keel along its back, is undoubtedly a development enabling it to cope with this problem, for hydrodynamically the pressure of water on the back of the fish as it lies passively on the bottom, forces it downwards against the river bed. The Humpback Sucker is a powerful fish of up to three feet in length and a weight of ten to fourteen pounds which has no difficulty in swimming against or across the current when occasion demands, but this artful modification of the body is an energy-saving device when the fish is inactive.

The catfishes are another dominant North American group. All the forms in fresh water between Canada and Guatemala, and there are at least twenty-four known species, belong to the family Ictaluridae. They all have four pairs of barbels around the mouth, scaleless, smooth skins, a rayless adipose fin on the back, and spines in front of the dorsal and pectoral fins. The largest species are the Flathead Catfish (*Pylodictis olivaris*), which is said to weigh up to 100 pounds, and grow to five and a half feet, and the Channel Catfish (*Ictalurus punctatus*) which grows to four feet and fifty-seven pounds and is a valuable food and sporting fish. Both were originally found only in the central United States, but the latter has been widely stocked for sport fishing outside its original range.

Many of the smaller catfishes are familiar as pond and aquarium fishes. The Brown Bullhead (*Ictalurus nebulosus*), originally found from Nova Scotia and Saskatchewan southwards to the Ohio Valley and Virginia, is now also found on the west coast as far north as British Columbia, and even on the Hawaiian Islands and in Europe. Other bullheads have been introduced as well, and they usually thrive when released, perhaps partly because they have the habit of guarding their eggs and the early young in dense shoals, so ensuring a high survival rate.

Another group of typically North American, nest-building fishes are the

sunfishes (family Centrarchidae). These are perch-like fishes with a very distinct, spiny dorsal fin, all carnivorous and many of them in favour as anglers' fishes. The Largemouth Black Bass (*Micropterus salmoides*) is one of the best known of these, growing up to a weight of twenty-five pounds, and partly because of its sporting qualities and size it has been widely introduced even to Europe. This is the giant of the family however, for several of the sunfishes never grow longer than four inches. One such is the Bluespotted Sunfish (*Enneacanthus gloriosus*) which is found through the eastern and southeastern United States. Another, the Everglades Pygmy Sunfish (*Elassoma evergladei*) is a veritable pygmy, sexually mature at one inch long and never growing longer than one and a half inches. Its habitat is very restricted even within the State of Florida. Another centrarchid with a restricted range is the Sacramento Perch (*Archoplites interruptus*) a two-foot native of California and the only member of the family west of the Rockies.

Left: large school of Grey Snappers (Lutjanus griseus), *common off Florida.*
Above: Black Bullhead (Ictalurus melas) *with its family. These small catfishes ensure a high survival rate by guarding their young in dense shoals.*
Top: the Nassau Grouper (Epinephelus striatus) *inhabits the Atlantic from North Carolina to Brazil, and reaches a length of 3–4 feet.*

The centrarchid sunfishes are very often beautifully coloured, and several of them have distinctively and brightly coloured fleshy lobes on the edge of the gill covers. They also have the habit of nest building; the male hollows out a small depression in the sand with his tail and entices a female into the nest. After egg-laying he jealously guards the nest, chasing away intruders and keeping the eggs clean by gently fanning them with his fins. The parental care continues for several days after the young have hatched.

Fishes which have taken to living in caves and have over their years in darkness gradually lost their well developed eyes are known from Africa, Asia, Australia and the Philippine Islands, but North America appears to have blind cave-fishes from more varied sources than other regions. The Mammoth Cave, Kentucky, is well known as the home of the Northern Cave-fish (*Amblyopsis spelaea*) which grows up to five inches in length, and in which the eyes are very much reduced and embedded under the skin. The body is a delicate, translucent, flesh pink. Another species, the

Southern Cave-fish (*Typhlichthys subterraneus*), lives in limestone caves in the south-eastern United States. The blind characin from Mexico, *Anoptichthys jordani*, which is found in subterranean streams and pools near San Luis Potosi in Mexico, is a very well known aquarium fish, perhaps partly from its strange pink colour with a silvery sheen, but more likely from the fact that it compensates for its complete lack of eyes by a frantic, restless activity in its search for food.

The fish fauna of North America is partly composed of groups which are widely dispersed across the northern landmasses of the world. A large part of the fauna, however, has evidently been present in the subcontinent for a very long time, as is shown by the many forms which have evolved only in North America, some of which have already been discussed. Interesting as these are they do not detract from the interest of the various archaic, primitive fishes found in North America today, many of which are known otherwise only from the fossil record.

One of these is the fascinating giant of the Mississippi system, the Paddlefish (*Polyodon spathula*) whose nearest and only close relative is found in the Yangtse River of China. The Paddlefish grows up to a maximum of six feet in length, and a weight of 168 pounds; it is smooth skinned, sturgeon-like in build with a long flattened snout, like a paddle about one third its total length. The Paddlefish is a plankton feeder, whose biology is relatively little known. It appears to have become much less abundant in recent years, and is now very restricted in its range due to pollution and the damming of many of the tributaries of the Mississippi.

If the Paddlefish looks rather like a long-snouted sturgeon, the Bowfin (*Amia calva*) has all the external appearances of an archaic fish. Its body is covered with heavy, overlapping bony scales, its head is protected by hard bony plates, there is a conspicuous bony gular plate on the throat, and the tail has the lopsided long upper lobe base of the primitive fishes. As a living fish it is unique to

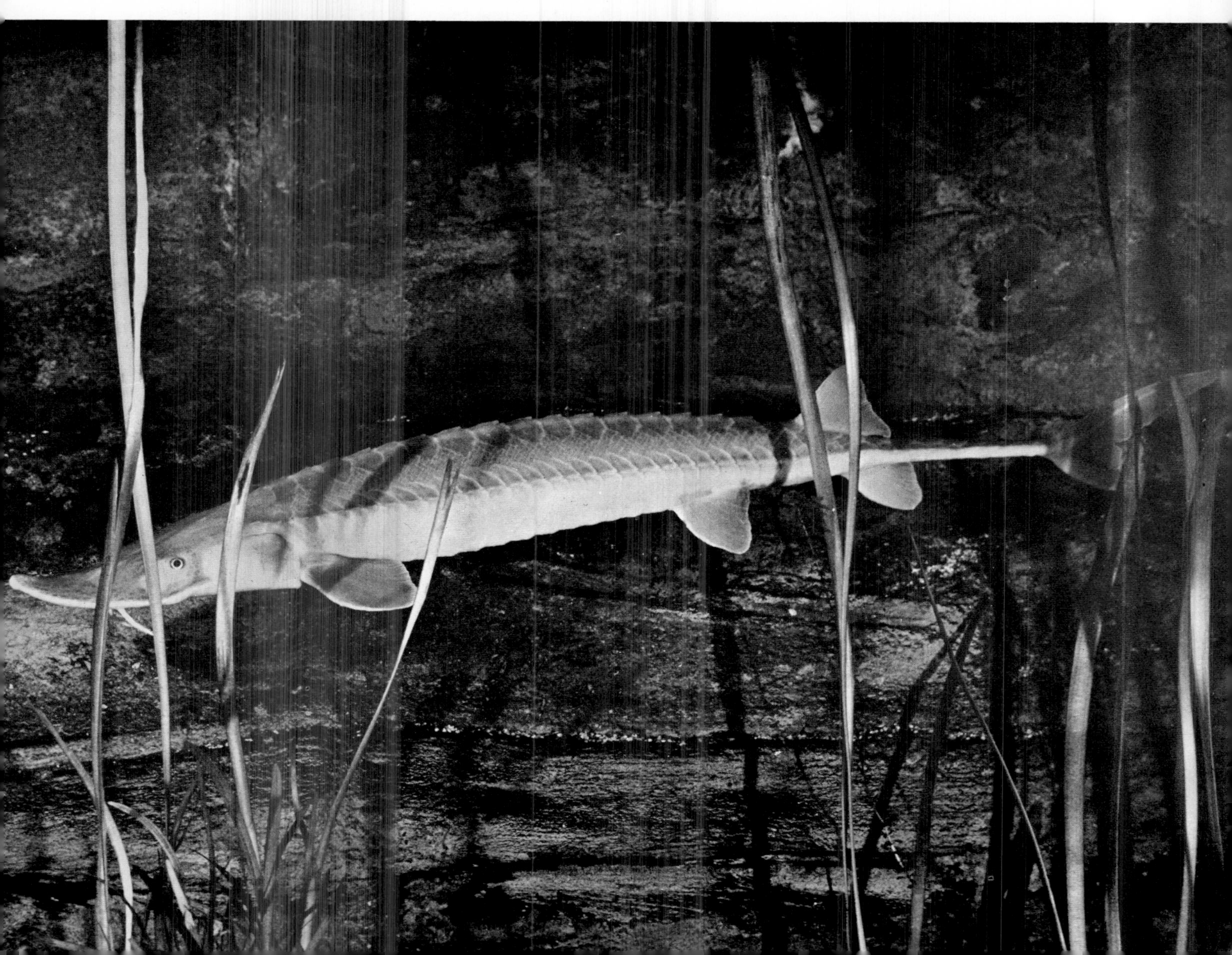

northern America, and is found from the Great Lakes and Mississippi region southwards to Florida and Texas. The Bowfin inhabits the shallow, still, weedy waters of bays, inlets and lagoons; it is often found in stagnant, oxygen-deficient water, and can use its large swim bladder, which has a cellular lining, as a lung, and breathe air at the surface. Using its swim bladder it can live for up to twenty-four hours out of water. It spawns in spring or early summer and the male clears an area in the weeds in ten to twelve inches of water, in which the female lays her eggs. The male guards the eggs, and although the young when hatched have an adhesive organ on the snout with which they cling to the vegetation, they are still shepherded and defended by the male. The Bowfin grows to a length of three feet, and being a relatively common fish and feeding on young fish and insects, can be quite a problem as predator and competitor with more sporting, or economically valuable fishes.

Eastern North America contains other heavily scaled, armour-plated archaic fishes in the gars (family Lepisosteidae). Similar fishes have been found in the Eocene deposits of Europe and India, and are perhaps fifty million years of age, so the five species found in America have a respectable and ancient lineage. The heavy scales, elongate bodies and long, pointed toothy snout give the gars a reptilian look. Most of the time they appear to hang motionless in the water. In fresh water they are found from south-eastern Canada to Panama, but they do not occur on the west of the Rocky Mountains. One or two species are also found in addition in the south in the sea, although usually keeping close to the coast, and most often staying in the vicinity of estuaries. The Alligator Gar (*Lepisosteus spatula*) is one of the sea-going gars, and is besides the largest of the northern species at ten feet long and a weight of 300 pounds. The most widely distributed of the gars is the Longnose Gar (*Lepisosteus osseus*) in which species the snout is particularly pronounced. It is found from the Great Lakes region south to the highlands of Mexico, usually in still, warm, open rivers and lakes. It spawns in spring, laying its greenish eggs on aquatic vegetation or on stony shoals. It grows to a length of about five feet, but due to its very narrow body it is relatively light, weighing about fourteen pounds.

Despite their interest to zoologists as virtual living fossils the gars are unappreciated in general. Anglers have little use for them as they rarely take a bait, and as large active predators they eat quantities of better fish. Commercial fishermen using nets find that they get their long snouts tangled in their nets, and although their flesh is edible it is little appreciated. Only the Seminole Indians of the Everglades region of Florida are said to eat gars where they are extremely abundant in the still swamps of this area.

The fishes of North America are a fascinating assemblage of endemic and archaic species, mixed with familiar fish of wider distribution and introduced forms from other parts of the world. As a fauna they show in many places only too well the effects of man's disturbance of their habitat. Some of the introductions have got out of hand and are now dominating the waters. In other places pollution has made rivers uninhabitable, and water conservators by building dams and weirs have decreased populations severely. Several American native fish are threatened with extinction from these causes. But the picture is not black throughout; northern American fishes are guarded by some of the most conservation-conscious fishery biologists in the world and many of the rarest of the threatened fishes have been closely studied. It is encouraging that such a varied and exciting fish fauna is so well known and protected.

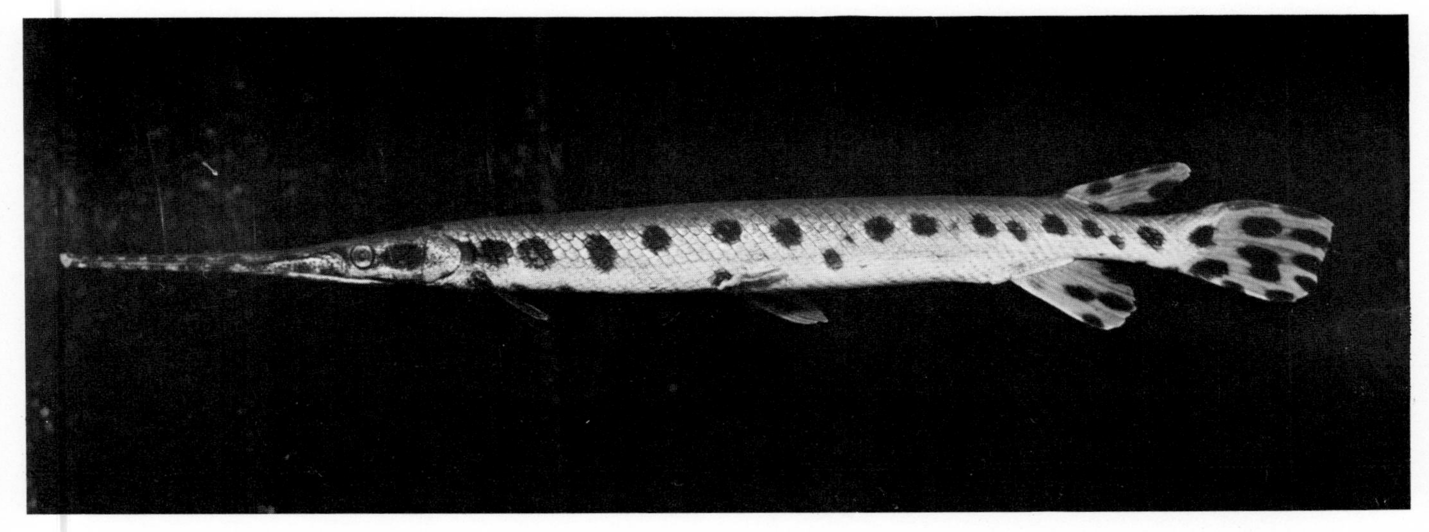

Left: Shovelnose Sturgeon (Scaphirhynchus platorynchus) *a fresh-water fish of central U.S.A. It lives on the bottom and locates its food by feeling with the sensitive barbels under its chin.*
Above: Common Garpike or Longnose Gar (Lepisosteus osseus).

BIRDS Bertel Bruun

The national bird of the United States is the Bald Eagle (*Haliaeetus leucocephalus*). In spite of its rather irreverent name it is a truly magnificent bird. Dark brown, with snow-white head and tail, its wings spread more than six feet as it glides along the water edge on the look out for suitable prey. Like most eagles, it does not shy away from the most foul-smelling carcase, and a large part of its food consists of dead fish washed ashore.

Like many other large raptors the world over, the Bald Eagle has suffered a drastic decline in numbers due to persecution by man. Besides this the species is threatened by the increasing and excessive use of insecticides which are accumulated in the uppermost part of the food chain, the carnivorous animals. Recent investigations have shown the Bald Eagle to be declining rapidly in numbers in spite of rigid protection. This is thought to be due to a marked decrease in fertility which is caused by the ingestion of sublethal dosages of insecticides. This decline has been noted even in Florida, a traditional stronghold of the species, and only about 3,000 birds are left. Much is being done to save this majestic bird but it is questionable if it is possible to avoid the extinction of the species in the southern part of its range.

All four species of loons, or divers as they are called in England, are found in the North American continent. Primarily northern in their distribution, loons can nevertheless be met with almost throughout the continent north of the Mexican border, and some venture even further south in the winter months. The most widespread and numerous is the Common Loon or Great Northern Diver (*Gavia immer*). In the north, its yodel-like call which ranges far and wide in the light summer night announces its presence in a nearby lake. Its large size and tasteful pattern of white and black makes it one of the most impressive of waterbirds. The Yellow-billed Loon or White-billed Diver (*Gavia adamsii*) is very similar and even larger but is restricted in range to the northwesternmost part of the country. The smaller Arctic Loon or Black-throated Diver (*Gavia arctica*) has a wider range but is in winter limited to the Pacific Coast whereas the Red-throated Loon (*Gavia stellata*) is met with commonly along both coasts in the cold season. Both it and the Common Loon will also winter on the Great Lakes, although they are more numerous along the coasts. All nest singly along the shores of lakes. In winter they are met with singly or in loose flocks.

Canada Goose (Branta canadensis).

Their close relatives the grebes are mainly birds of ponds and lakes, although most have a tendency to move to the coasts in winter. The largest, the Western Grebe (*Aechmophorus occidentalis*), is famed for its elaborate courtship display which is seen in spring on the western lakes where it breeds in colonies. Side by side a pair will run along the water with wings half extended and their snake-like necks bent in a graceful arch. The Horned or Slavonian Grebe (*Podiceps auritus*) is the most numerous grebe, except in the south-west. It is much smaller than the Western Grebe. In winter it is often seen in harbours, often several together, and it can here be studied at close range. The similar Eared or Black-necked Grebe (*Podiceps nigricollis*) also has a western distribution and breeds in colonies. The nest is placed in floating vegetation, as are those of the other grebes. The Pied-billed Grebe (*Podilymbus podiceps*) with its dull brown colours and short heavy bill is the most widespread, and can be met with in shallow water almost anywhere throughout the continent with the exception of the most northern part. It leads a retiring life and is very prone to submerge gradually leaving only its head above the water if it feels danger approaching. This habit is shared by the tiny Least Grebe (*Podiceps dominicus*) of the swamps and lakes of the southernmost part of the continent.

Bordered by the Pacific in the west and the Atlantic on the east, North

America naturally has its share of pelagic species. Foremost among these and most truly pelagic are the tubinares. Perfectly adapted to a life at sea these birds only rarely get within sight of land and only at breeding time do they go there voluntarily. On the Atlantic side the most numerous species is Wilson's Petrel (*Oceanites oceanicus*) which might even be the most numerous bird in the world. Wilson's Petrel does not breed in North America but is a visitor from the Antarctic region where it nests. In early summer it starts appearing off the Atlantic coast and soon builds up in numbers. By the hundreds these agile little birds with their butterfly-like flight will skim the waves on the look out for planktonic animals and often they will surround fishing boats in the hope of picking up titbits. As summer progresses they move northwards and soon they start crossing the Atlantic as they continue their large figure of eight migration which eventually leads them back to the Antarctic waters.

A similar type of migration is performed by the Greater Shearwater (*Puffinus gravis*) and the Sooty Shearwater (*Puffinus griseus*), the latter being more numerously found in the Pacific. Here the Pink-footed Shearwater (*Puffinus creatopus*), Manx Shearwater (*Puffinus puffinus*) and Slender-billed Shearwater (*Puffinus tenuirostris*) are also found, only the Manx as a breeding bird. The Cahow (*Pterodroma cahow*) is a small shearwater for long thought extinct until rediscovered as a breeding bird in Bermuda in recent years. Several petrels nest along the Pacific, most numerously the light-coloured Fork-tailed Petrel (*Oceanodroma furcata*). The only species nesting on the Atlantic side is Leach's Petrel (*Oceanodroma leucorhoa*) which is also found on the west coast. The same is true of the larger and more northerly distributed Fulmar (*Fulmarus glacialis*). The largest of the tubinares, the albatrosses, are primarily birds of the southern hemisphere and the only member of this majestic family to occur regularly off the North American coast is the Black-footed Albatross (*Diomedea nigripes*) which is found far out to sea in the Pacific.

Besides the albatrosses other tropical pelagic species are found off the

Left: Horned Grebe (Podiceps auritus) *scaring away an intruder from its nest.*
Below: adult and young Western Grebe (Aechmophorus occidentalis).
Bottom: Common Loon (Gavia immer).
Above: Roseate Spoonbill (Ajaia ajaja).
Right: Roseate Spoonbills and Common Egrets (Casmerodius albus).

southern part of the west coast and in the West Indies. Among the most spectacular is the Magnificent Frigate-bird or Man-of-War Bird (*Fregata magnificens*). These masters of the air have very long pointed wings and a long forked tail. They are extremely graceful on the wing but largely make their living as robbers of gulls, terns and pelicans, forcing them to give up their hard-earned prey by constantly pursuing and pestering them. Even tropic-birds, pigeon-sized pelagic relatives of the frigate-birds found in the same region although usually further out to sea, are not left alone. The frigate-birds live in colonies on small islands where they place their nests in small trees and bushes. The male can inflate its red throat-pouch to an enormous size. On the breeding grounds the frigate-birds share the colonies with many other species, among them the Brown Pelican (*Pelecanus occidentalis*) which is also found further north along the Gulf Coast and the west and east coasts. Unlike other pelicans such as the White Pelican (*Pelecanus erythrorhynchos*) which breeds in the north-west, wintering along the west coast and the Gulf Coast and obtaining its food by shovelling fish with its large bill, the Brown Pelican fishes by diving. It flies at a medium height over the water, and when a fish is spotted it plunges into the sea head-first in tern fashion and more often than not is successful in obtaining its prey. The same manner of fishing is used by the Gannet (*Sula bassana*) which breeds in large colonies on the northern Atlantic coast, and by its tropical relatives the boobies found in the tropical seas. Cormorants, of which there are several species in North America, also live on fish, but they pursue them underwater and grasp them with their bills. The Anhinga (*Anhinga anhinga*) is another relative which pursues fish. It looks somewhat like a cormorant, but has a very long slender neck and a very pointed bill on which it impales the fish. It is a bird of marshes and swamps, as opposed to the cormorants which are mainly found along sea shores and in larger lakes.

The Double-crested Cormorant (*Phalacrocorax auritus*) is the most widespread of the cormorants, and in the south often shares its colonies with the Brown Pelicans. Other inhabitants of these often very large mixed colonies are the ibises. The most spectacular of these is the Scarlet Ibis (*Eudocimus ruber*) found in South America and recently introduced in southernmost Florida. Further north the White Ibis (*Eudocimus albus*), Glossy Ibis (*Plegadis falcinellus*) and White-faced Ibis (*Plegadis chihi*) extend their range. The latter two species are very similar but the Glossy Ibis has an eastern, the White-faced a western distribution. Along the Gulf Coast another exotic but much rarer bird may be encountered, the Roseate Spoonbill (*Ajaia ajaja*). With its spatulate bill and mainly pink plumage this remarkable species is one of the most exotic birds of the North American avifauna. At one time extremely rare, the species is now slowly regaining its former numbers. The American or Greater Flamingo (*Phoenicopterus ruber*) is restricted to the West Indies where its colonies are found on mudflats. Although most flamingos move about in large flocks, occasional stragglers of this beautiful species reach the southern coast of the United States. The only North American stork is the Wood Ibis (*Mycteria americana*) which is found in southern swamps and marshes. It nests in colonies in trees, often tall cypresses.

Sharing the colonies of Brown Pelicans, Double-crested Cormorants, and ibises are several species of herons. Actually they often dominate these colonies and in the northern parts of the continent the colonies might consist exclusively of herons. Remarkable is the number of white herons met with in North America. The largest of these is the Great White Heron (*Ardea occidentalis*) which is limited in range to the southernmost part of Florida. Much more widespread is the Common or Great White Egret (*Casmerodius albus*) and the Snowy Egret (*Leucophoyx thula*). Both these species suffered a tremendous decline during the nineteenth and the beginning of the twentieth century due to hunting. Their aigrettes were much prized and the populations were decimated by the millinery industry. Thanks to rigid protection and perhaps even more to

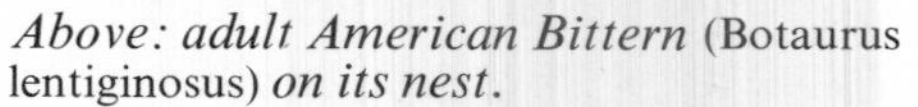

Above: adult American Bittern (Botaurus lentiginosus) *on its nest.*
Top: a young Yellow-crowned Night Heron (Nyctanassa violacea) *just growing its adult plumage.*
Above right: the Wood Ibis (Mycteria americana) *is found locally in swamps and marshes of the southern U.S. and Mexico.*
Right: Double-crested Cormorant (Phalacrocorax auritus).
Far right: a colony of Wood Ibises.

a change in fashion, the egrets are once again common. A newcomer to North America is the Cattle Egret (*Bubulcus ibis*). In the 1930s the species crossed from Africa to Guyana where it established itself in small numbers and slowly started spreading. In the 1950s it reached the North American continent and since then has spread with an explosive speed. The most widespread heron is the majestic Great Blue Heron (*Ardea herodias*) which can be met with almost anywhere on the continent. Two other widespread species of the heron family are the Black-crowned Night Heron (*Nycticorax nycticorax*) and the American Bittern (*Botaurus lentiginosus*). Both of these are mainly nocturnal in habits although the Night Heron is often encountered during the daylight hours.

Waterfowl are prominent among the birds of North America. Most numerous of the geese and most widely distributed of the waterfowl is the Canada Goose (*Branta canadensis*) which has also been introduced into Europe where it has established itself in England and Sweden. In North America, it is found breeding north to the Arctic Sea and south through the northern part of the United States. No less than six sub-species are recognized, varying from the Giant Canada Goose (*Branta canadensis maxima*) to the Cackling Goose (*Branta canadensis minima*). The sight of Canada Geese on the move in autumn and spring is an integral part of the American natural scene and marks more than anything else the changing seasons. Canada Geese usually move about in flocks, honking as they fly, and on longer travels assume the V-formation so characteristic of waterfowl. The Canada Geese usually nest on the ground, never far from water but, unlike the several other geese, not in colonies.

Two close relatives, the Brant or Brent Goose (*Branta bernicla*) and the darker Black Brant (*Branta nigricans*) are examples of such colony nesters. In summer they are found only in the northernmost part of Canada and Alaska, but in winter they move south along both coasts where they can be met with in very large flocks. The Brant suffered a tremendous decline in numbers some years ago when the eel-grass on which it lives disappeared from many areas along the Atlantic coast. In recent years the eel-grass has returned, and with it the Brant. The Black Brant, which is restricted to the Pacific coast, did not suffer as the eel-grass was not affected there. Both species are strictly coastal in habitat.

Most spectacular of the geese is the Snow Goose (*Chen hyperborea*) with its snow-white plumage accentuated by the pitch black wing tips. The Snow Geese nest in northernmost Canada and Alaska on islands in tundra lakes in large colonies. When autumn comes the geese leave the breeding grounds in large flocks, whole populations moving at the same time. On migration they will often cover hundreds of miles in one long flight and then settle for longer periods at favoured haunts en route before reaching their wintering grounds along the coast of the Atlantic, Mexican Gulf and Pacific Ocean. A small relative, the duck-sized Ross Goose (*Chen rossii*) has a very limited range. The White-fronted Goose (*Anser albifrons*) is mainly western in its distribution.

As in Europe many city parks harbour Mute Swans (*Cygnus olor*) but in the eastern part of the U.S.A. this species is now established in the wild. The two native American swans are the Whistling Swan (*Olor columbianus*) and the Trumpeter Swan (*Olor buccinator*). The Whistling Swan is quite common. It breeds in the far north, wintering in large flocks on lakes and along the coast. The quite similar but somewhat larger Trumpeter Swan has been on the verge of extinction. By the 1930s there were only a few dozen Trumpeters left, almost all found in the spectacular Yellowstone Park. Since then rigid protection has caused the population to grow steadily and the species is now rapidly extending its range in the north-west, regaining lost ground. It is a triumph for American conservationists and an encouragement in the attempt to save other species threatened by man's ever-increasing encroachment on nature.

The hunting of waterfowl is regulated in North America by the Migratory Bird Act, a treaty between the United States, Canada and Mexico. Although only applying to migratory birds, almost all waterfowl fall in that category making it possible for the government to issue and enforce game laws restricting the hunting of rare

species and species which have suffered from overshooting. The regulations are changed from year to year based on information obtained through census during and outside the breeding season. An estimate of the shooting and of the breeding stock enables the authorities to take measures to protect certain species before too serious a decline has taken place. The protection is further enhanced by many wildlife refuges operated by Federal and State agencies as well as by private organizations.

The most widely distributed duck is the Pintail (*Anas acuta*) which is abundant in the west but also common in the east. In winter it often mixes with other ducks like the Mallard (*Anas platyrhynchos*), most common in the west, and the Black Duck (*Anas rubripes*), abundant in the east. The Baldpate or American Wigeon (*Anas americana*) is a very common species. Its soft whistle is very characteristic and its white wing patches also help in identification. Its close relative, the European Wigeon (*Anas penelope*) is a regular visitor to the Atlantic seaboard in winter. It comes from its breeding grounds in Iceland and several birds ringed there have later been recovered in North America. The Shoveller (*Spatula clypeata*) and the Gadwall (*Anas strepera*) are species shared with the palaearctic. The Green-winged Teal (*Anas carolinensis*) is the smallest North American duck. Like most of the ducks, its main breeding range is the west and north-west. Here in spring the many prairie lakes are teeming with waterfowl of many species, among them the shy Blue-winged Teal (*Anas discors*). Further south this species is replaced by the quite similar Cinnamon Teal (*Anas cyanoptera*).

A most beautiful and exotic-looking waterfowl is the widespread Wood Duck (*Aix sponsa*), which seeks out secluded lakes surrounded by woods. Its nest is placed in a cavity of a tree and it is able to dodge agilely among the trees, unlike other ducks whose fast flight is direct and without much manoeuvrability. It is not related at all to the tree ducks of which two species, the Fulvous (*Dendrocygna bicolor*) and the Black-bellied (*Dendrocygna autumnalis*), inhabit Mexico and the southernmost United States. They are shy, retiring birds coming into the open only at night when they seek their food in rice and corn fields. They call with a shrill whistle.

The bay ducks are all found nesting in the north-western part of the continent. The Greater Scaup (*Aythya marila*) is most northerly in its distribution. The very similar Lesser Scaup (*Aythya affinis*) is the most abundant, and the red-headed Canvasback (*Aythya valisineria*) with its very long, sloping bill is a prized game bird. In winter they are found along all the coasts and also often in harbours where they can be studied at close range. The spectacularly coloured Harlequin Duck (*Histrionicus histrionicus*) is found in the north where it inhabits the swiftest running rivers; it moves with great agility in the most turbulent waters and even the downy ducklings manoeuvre easily in rushing waters.

When travelling in the southern part of the continent one usually encounters vultures. The most widespread is the Turkey Vulture (*Cathartes aura*) which is black with a naked red head, as

A flock of Canada Geese (Branta canadensis). *Some are hunting for food in the mud at the bottom of the lake.*

opposed to the Black Vulture (*Coragyps atratus*) which has a black head. Both frequent the roadsides where they patiently await an animal falling prey to the ever-increasing traffic. The vultures themselves are sometimes killed by a car as they move away from a carcase too slowly. The Turkey Vulture is often seen soaring very high and at a distance looks quite impressive. When seen at short range, assembled around a carcase, the vultures certainly look sinister, but their

Right: Snow Geese (Chen hyperborea) *breed in the far north and migrate as far south as Mexico for the winter. Below right: Snow Geese in their typical formation flight.*

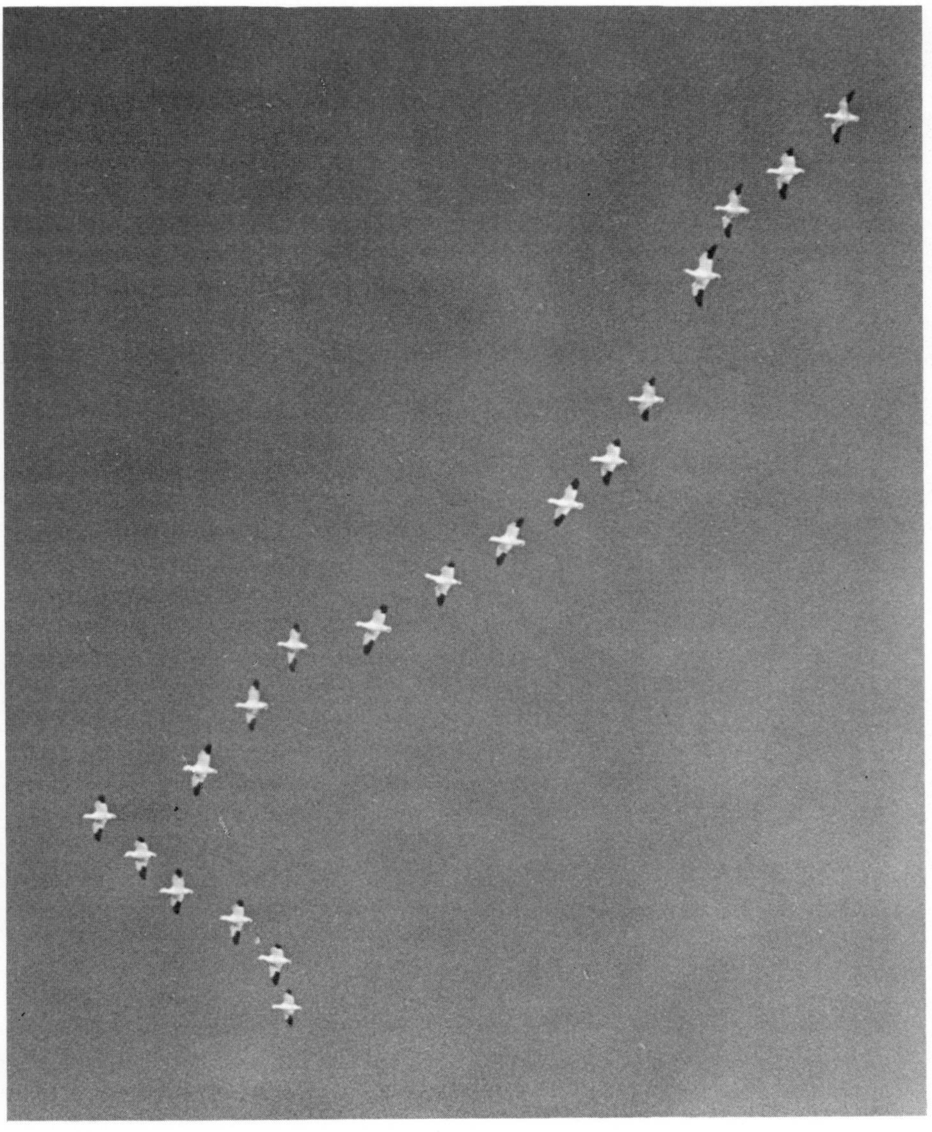

Left: male Wood or Caroline Duck (Aix sponsa) *in breeding plumage.*
Below left: Trumpeter Swans (Olor buccinator) *were recently close to extinction but are now increasing in game reserves in the Rocky Mountains.*
Below: a group of Pintails (Anas acuta).

service as scavengers is considerable. Often associating with the vulture is the Caracara (*Caracara cheriway*), a long-legged scavenger of the southern prairies. It has a very large bill and its beautiful black and white plumage makes it a pleasant contrast to the more common but much uglier Black and Turkey Vultures.

The famous California Condor (*Gymnogyps californianus*) is almost extinct, only a few pairs holding their own in a small area in southern California. Its wing span of ten feet makes it one of the most impressive birds. Common in the tropical forest region is the King Vulture (*Sarcorhamphus papa*), a magnificent black and white bird with a bizarre and colourful pattern to its naked head.

By far the most numerous group of birds of prey are the buteos. No less than thirteen are found in the United States alone. The most widespread is the Red-tailed Hawk (*Buteo jamaicensis*) which can be met with throughout the continent with the exception of the most northern parts where the Rough-legged Hawk (*Buteo lagopus*) breeds. This species, like several other buteos, occurs in two colour phases, a light and a dark, the latter almost completely black. The best place to observe autumn migration of raptors in North America is undoubtedly the appropriately named Hawk Mountain in Pennsylvania. Here thousands of hawks and many other birds of prey can be seen moving south along the mountain ranges.

Occasionally a Golden Eagle (*Aquila chrysaëtos*) may also be seen here although this magnificent species is more common and widespread in the western part of the continent. Falcons might also pass by here although they are less bound to the upward current created along mountain ranges and coastlines than are the broad-winged raptors. The Sparrow Hawk or American Kestrel (*Falco sparverius*) is a most charming and beautiful little falcon found throughout most of the continent. As in Europe the Peregrine Falcon (*Falco peregrinus*) has suffered an almost catastrophic decline in numbers in recent years whereas the Gyrfalcon (*Falco gyrfalco*) in the virtually uninhabited north has been holding its own. The most brightly coloured of the American falcons is the Aplomado Falcon (*Falco femoralis*), a black, white and red species found in the southernmost parts.

A very widespread bird of prey is the Osprey (*Pandion haliaetus*), a magnificent black and white species which

lives solely on fish which it catches as it plunges into the water, its sharp talons outstretched to grasp its slippery prey. It builds a large bulky nest which is usually placed in trees or even telegraph poles, but occasionally on the ground.

Birds of prey have been persecuted because of their competition with man for the same game. This competition has proved to be non-existent for all practical purposes, as in most cases the game falling prey to raptors are unhealthy individuals which would have succumbed anyway. Besides, the inroads made by the birds of prey on a given species are negligible compared to the mass destruction wrought by man. Prominent among the game we thought we defended are the gallinaceous birds. This family of large to medium-sized birds is found throughout the continent and in many areas comprises a prominent part of the avifauna.

In the far north, on the tundras and in the mountains are found the ptarmigans, the most widespread being the Willow Ptarmigan (*Lagopus lagopus*) and the Rock Ptarmigan (*Lagopus mutus*). The White-tailed Ptarmigan (*Lagopus leucurus*) is limited to the western mountains, and is completely white in winter. In the vast belt of coniferous forests, stretching across the northern part of the continent from the Atlantic to the Pacific, Spruce Grouse (*Canachites canadensis*) and Ruffed Grouse (*Bonasa umbellus*) are common. The latter attracts attention by the 'drumming' sound it makes by rapidly beating the air with its wings during its picturesque display.

The prairies are inhabited by various grouse and in the southern part of the continent are found a large number of different quails. Foremost among the American gallinaceous birds is the Turkey (*Meleagris gallopavo*). The wild form resembles the domestic but is somewhat slimmer. It is a bird of open woodland, and although it has suffered much from hunting it is still found in many areas.

In the wooded swamps of the south is found the strange Limpkin (*Aramus guarauna*), a brown, speckled, heron-sized bird with a long slightly decurved bill with which it picks up and eats large snails, its staple food. More picturesque is the Sunbittern (*Eurypygidae helias*), also an inhabitant of the tropical swamps, which is famed for its courtship when it displays its beautifully patterned wings and tail. Less retiring in habits and more often found in the open is the American Jacana (*Jacana spinosa*), a gallinule-like bird with tremendously long toes which enable it to walk with ease on floating vegetation.

The Purple Gallinule (*Porphyrula martinica*) is another marsh dweller of great beauty with its metallic purple colours. Further north the swamps are inhabited by a number of different rails, varying in size from the tiny four-and-a-half-inch Black Rail (*Laterallus jamaicensis*) to the fourteen-inch King Rail (*Rallus elegans*). The American Coot (*Fulica americana*) is widespread and much more often seen than the secretive rails. Several rails inhabiting islands in the West Indies have become extinct after the introduction of cats and rats.

Another and more spectacular North American bird threatened with extinction is the stately Whooping Crane (*Grus americana*). As opposed to the common and widespread Sandhill Crane (*Grus canadensis*) the Whooping Crane has never been very numerous, but in the present century its population has dropped to about forty individuals. In spite of rigid protection both of its breeding grounds in north-western Canada and wintering grounds

on the Gulf Coast the species is barely holding its own. Eight individuals are now kept in captivity in an attempt to breed them for later release. The outcome of the tremendous effort to save this species is still in doubt, but effort is followed with great interest by and receives widespread support from the American public.

The Eskimo Curlew (*Numenius borealis*) was long thought to be extinct, but in 1959 was rediscovered on the Gulf Coast on spring migration where a few individuals have since been seen each spring. The species was formerly common. It bred on the tundra of northern Canada and in autumn migrated across the Atlantic to South America where it spent the winter on the Argentine pampas. The spring migration was inland through the prairie States where thousands were killed on their way to the nesting grounds. This interesting migratory route is today followed by many North American shorebirds, notably the American Golden Plover (*Pluvialis dominica*) which covers the distance of more than 1,000 miles from Labrador to South America in one uninterrupted flight. An even more impressive flight is undertaken by the Bristle-thighed Curlew (*Numenius tahitiensis*) which migrates directly from its breeding grounds in Alaska to its wintering grounds on Hawaii.

The Arctic offers a very large and rich area for nesting in the short but lush summer. Thousands of shorebirds take advantage of this, foremost among them the so-called peeps, a group of small sandpipers. Most abundant of these is the small Semipalmated Sandpiper (*Ereunetes pusillus*), but very common also are the similar Least Sandpiper (*Erolia minutilla*) and Western Sandpiper (*Ereunetes mauri*). Outside the breeding season they are encountered in flocks on mudbanks along the shores of the open sea, marshes, lakes and rivers. The winter is spent in the southern part of North America or in South America.

In the northern coniferous belt are found a large number of other shorebirds breeding along the many lakes and swamps. Most common is the Greater Yellowlegs (*Totanus melanoleucus*) and the very similar Lesser Yellowlegs (*Totanus flavipes*). Both are very long-legged, graceful birds with long, slender bills. They are common throughout the continent at the time of migration and spend the winter in the southern part of the continent. Several other species inhabit this region but are less numerous. On the western prairies with their many lakes and marshes large species like the well-named Long-billed Curlew (*Numenius americanus*) and the Marbled Godwit (*Limosa fedoa*), and the most beautiful of American shorebirds, the gracious American Avocet (*Recurvirostra americana*) with its needle-thin upturned bill, are found nesting. The elegant Wilson's Phalarope (*Steganopus tricolor*) is also indigenous to that area.

Far left: Turkey Vultures (Cathartes aura).
Above: Whooping Crane (Grus americana) *defending its territory.*
Left: Sandhill Crane (Grus canadensis).

On the coasts the breeding shorebirds vary from area to area and there is also quite a difference between the species wintering along the Pacific and the Atlantic coasts. On the Pacific side in summer are found Snowy Plovers (*Charadrius alexandrinus*) and the magnificent Black Oystercatcher (*Haematopus bachmani*) whereas on the Atlantic side the American Oystercatcher (*Haematopus palliatus*), Piping Plover (*Charadrius nulodus*) and Wilson's Plover (*Charadrius wilsonia*) breed.

Most widespread of the North American shorebirds are the Spotted Sandpiper (*Actitis macularia*), a small very active bird mainly found breeding along rivers and lakeshores, the Killdeer (*Charadrius vociferus*), a large banded plover common on fields and pastures, and the Snipe (*Capella gallinago*) of marshes and bogs. In the south the extremely long-legged, elegantly black and white coloured Black-necked Stilt (*Himantopus mexicanus*) is common in marshes and swamps.

An even stranger looking inhabitant of the southern shores is the Black Skimmer (*Rynchops nigra*). It is a large tern-like bird, black above and white below, and has a long coral red bill with the mandible (lower part) much longer than the maxilla (upper part). It obtains its food by flying low over the water with the mandible dipped into the water stirring up small aquatic life. It nests in colonies on sandy seashores, often in company with Common Terns (*Sterna hirundo*), another species common on the Atlantic shore. Several other terns are found in North America ranging in size from the Little or Least Tern (*Sterna albifrons*) to the Caspian Tern (*Hydroprogne caspia*) and in habitat from the marsh dwelling Black Tern (*Chlidonias niger*) to the pelagic Noddy Tern (*Anoüs stolidus*).

Although terns are commonly seen along the seashores, gulls are much more evident here, as they are also in

Left: a flight of Sandhill Cranes (Grus canadensis) *over Alaska.*
Right: Bald Eagle (Haliaeetus leucocephalus) *with the remains of its prey.*

harbours and on garbage dumps, particularly the Herring Gull (*Larus argentatus*) which can be met with all over the continent. Even bigger in size and with solid black back and upper wings are the Western Gull (*Larus occidentalis*) of the Pacific coast and the Greater Black-backed Gull (*Larus marinus*) of the Atlantic, where the species has been extending its range to the south in recent years.

Most gulls do not venture far out to sea, but there are exceptions to this. Most pelagic of the gulls is probably the Kittiwake or Blacklegged Kittiwake (*Rissa tridactyla*). This is a very common breeder on the northern cliffs where it is frequently found associated with Razorbills (*Alca torda*) and Common Murre or Guillemots (*Uria aalge*). These and several other species of alcids are found in the North Atlantic, but the group is much better represented in the northern Pacific where no less than sixteen different species are found. Most spectacular among these is the Tufted Puffin (*Lunda cirrhata*) which is large and heavy with an enormous red bill, long straw-coloured tufts and generally black colours.

The alcids are expert fishers, pursuing their prey while swimming swiftly underwater. A completely different mode of fishing is practised by the kingfishers. These birds are usually brightly coloured and have dispropor-

American Avocets (Recurvirostra americana) *and Black-necked Stilts* (Himantopus mexicanus) *are long-legged wading birds. The Avocet's bill curves upwards and it has a reddish head.*

tionately large heads and bills. They fish by throwing themselves headlong into the water. The larger species such as the Belted Kingfisher (*Megaceryle alcyon*), which is common north of the Mexican border, often hover over the water in the same fashion as terns. On the other hand, the Pygmy Kingfisher (*Chloroceryle alma*) of tropical streams lives mainly on insects it catches in the air, and the Green Kingfisher (*Chloroceryle americana*) usually waits patiently for fish to come near from a branch overhanging the water.

Only two species of parrots have occurred north of the Mexican border, the Thick-billed Parrot (*Rhynchopsitta pachyrhyncha*), which has recently been found in very small numbers just north of the border, and the Carolina Parakeet (*Conuropsis carolinensis*), a small green parrot which used to roam the cypress forests of the south-east but which has apparently been extinct since about 1914.

Hummingbirds are limited in their distribution to the New World. Brilliantly coloured as many are, they have been described as a 'glittering fragment of the Rainbow' by the famous painter and ornithologist John James Audubon. Their iridescent colours are of a different nature from those of most other birds, which have pigments in their feathers. The hummingbirds' colours are produced by the texture of the feathers, much as colours are produced by the facets of a diamond, and there is little or no pigment to account for the brilliance. Among the hummingbirds are found the smallest of birds; the Bee Humming-

Far left: a Great Blue Heron (Ardea herodias) *surveying the scene from a pine tree.*
Top left: Snowy Egrets (Leucophoyx thula) *with their nests.*
Left: Belted Kingfisher (Megaceryle alcyon) *diving for a fish.*
Above: the snow white winter plumage of the Willow Ptarmigan (Lagopus lagopus) *is an excellent camouflage.*

bird (*Mellisuga helenae*) found on Cuba is the smallest bird with a total length of only two and a half inches. Most unique though is the flight of hummingbirds. Although several other birds can hover for short moments, the hummingbirds do this much more frequently than other species, and they are able to fly backwards which no other bird can do.

The hummingbird uses its wings in such a way that it obtains an updrift with both the downward and upward stroke of the wing, whereas other birds only do it on the downstroke. Although the hummingbird beats its wings about seventy-five times per second, for a bird of its size this is not a particularly fast rate; on the contrary it is possibly rather slow.

The small size and very high metabolic rate of the hummingbirds require a large amount of energy, and it has been calculated that a hummingbird eats about half its own weight each day in almost pure sugar. This is obtained from flowers, into which the hovering bird dips its long tubular tongue. The diet is supplemented by a few small insects. At night, when food cannot be obtained and the metabolic rate therefore not maintained at the same high level the bird goes into a semicomatose state somewhat similar to that of hibernating animals.

In eastern North America only one species is found, the Ruby-throated Hummingbird (*Archilochus colubris*). It reaches as far north as north of the Canadian border. It is migratory and is even capable of crossing the Gulf of Mexico. The species reaching the furthest north is the Rufous Hummingbird (*Selasphorus rufus*) which is found breeding from California to southern Alaska. In the west many more species are found and in Central America more than a hundred are counted.

The hummingbirds are most closely related to the swifts, another family of aero-acrobats. Swifts have specialized in speed more than manoeuvrability though, and are, compared to the hummingbirds, as jet planes compared to helicopters.

Below: group of Common Puffins (Fratercula arctica) *in summer plumage. Right: Nighthawk* (Chordeiles minor). *These birds are not in fact hawks, but are related to the Whip-poor-will* (Caprimulgus vociferus).

Far left top: part of a flock of Laughing Gulls (Larus atricilla) *in breeding plumage.*
Far left bottom: an argument between neighbouring Elegant Terns (Thalasseus elegans) *while the egg lies unnoticed.*
Left: a large colony of Elegant Terns.
Below: Herring Gulls (Larus argentatus).

Left: the Barn Owl (Tyto alba) *hunts at night and preys on mice and rats.*
Above: adult Barn Owl (Tyto alba) *with young; they build their nests in barns, old buildings and in holes in trees.*
Below: the diurnal Snowy Owl (Nyctea scandiaca).

The only member of the family found in the east is the Chimney Swift (*Chaetura pelagica*), so called for its habit of roosting in flocks in tall chimneys when on migration. Several other species are found in the west and in central America, among them the beautifully black and white coloured White-throated Swift (*Aëronautes saxatalis*). Almost swift-like in silhouette are the much larger goatsuckers, a group of long-winged, nocturnal birds which catch insects in flight in their very wide gape. There are several species in North America of which the most commonly seen (because of its partially diurnal habits) is the Nighthawk (*Chordeiles minor*). With the exception of the northernmost parts it is found nesting throughout the continent. Several of the species have very characteristic calls which are often imitated in their names. Such is the case of the Whip-poor-will (*Caprimulgus vociferus*) and Chuck-will's-widow (*Caprimulgus carolinensis*), both of which are commonly heard in the summer nights in the east.

The Poor-will (*Phalaenoptilus nuttallii*), common in the west, is the only

Far left top: female Ruby-throated Hummingbird (Archilochus colubris) *feeding her young which is almost too large for its mud nest.*
Far left bottom: a male Ruby-throated Hummingbird hovering near a flower while sipping nectar.
Left: male Gilded Flicker (Colaptes chrysoides) *perched on a giant cactus.*
Below: Saw-whet Owl (Aegolius acadica) *perched in a maple tree.*

bird which is known to have hibernated, as a bird was found in such a condition in several successive winters hidden in a small cave.

Sharing the nocturnal habits of the goatsuckers are the owls. In the Arctic the beautiful Snowy Owl (*Nyctea scandiaca*) has its home. Its main food is the lemming and its abundance varies with that of its prey. In some winters the Snowy Owls move quite far south when their food supply in the north is poor. In the northern coniferous woods is found the enormous Great Grey Owl (*Strix nebulosa*) which is even larger than the much more widespread Great Horned Owl (*Bubo virginianus*). Both are ferocious hunters living mainly on rabbits and other rodents. The Great Horned Owl is also known to take a number of birds, among them the only slightly smaller but much less aggressive Barred Owl (*Strix varia*), common in southern swamps.

Further south, and often near human habitations are found the Screech Owl (*Otus asio*) which is 'eared', and the larger and more long-legged Barn Owl (*Tyto alba*). On the western prairies and plains is found the Burrowing Owl (*Speotyto cunicularia*), a medium-sized owl with a very short tail and long legs. It is often found in prairie dog 'towns' where it uses an abandoned hole for a nesting site. Much of the day is spent in front of the hole, and the bird has a habit of turning its head in a full circle as it surveys the surroundings. It also often closes one eye at a time, and frequently does fast knee bends. In the south-western desert the smallest of the owls, the Elf Owl (*Micrathene whitneyi*) uses holes in the giant cactus for a nesting site. One of the most widespread birds in North America is the Mourning Dove (*Zenaidura macroura*), a small long-tailed dove found from the Canadian border southwards. It is very common in cultivated areas and has benefited much from the clearing of land, a process so detrimental to many other species. It is usually seen in pairs and is often quite tame. Its soft cooing is a pleasant though sad-sounding call which has earned it its name. It is a favoured game-bird in many areas, and more than twenty million are killed by hunters every year. It is interesting that this species has been able to replace such losses, whereas its relative, the Passenger Pigeon (*Ectopistes migratorius*), which was probably even more numerous, became extinct in 1914 when the last specimen died in Cincinatti Zoo. The Passenger Pigeon was highly social and its nesting success was dependent on colonies numbering thousands of pairs. This kind of concentration during and outside the breeding season made the birds most vulnerable to mass slaughter and as soon as the population was brought below a certain level the birds were unable to breed successfully as they lacked the social stimulus of their former hordes.

The North American cuckoos do not share the interesting habit of nest parasitism with their Old World relatives. The Yellow-billed Cuckoo (*Coccyzus americanus*) is the most widespread species. It is a long-tailed, rather sluggish bird of forest and brush and usually keeps well hidden.

The cuckoos have some interesting relatives. Best known is the very long-tailed Roadrunner (*Geococcyx californianus*) found in the arid parts of the south-west. It rarely flies but runs very fast on its long legs, and it often raises its tail and crest. Its food consists of snakes, lizards and insects. The anis, of which there are two species, the Groove-billed (*Crotophaga sulcirostris*) and the Smooth-billed *Crotophaga ani*) are very long-tailed black birds with very large bills. Several females often share the same nest; they separate the different layers of eggs with leaves which are removed as the top layer hatches. Incubation is shared by the females. The anis are found in the southernmost part of the continent.

Much of North America is wooded, and as in other parts of the world woodpeckers are common inhabitants of these forested areas.

Most widespread of the woodpeckers are the Hairy (*Dendrocopos villosus*) and the similar but smaller Downy Woodpecker (*Dendrocopos pubescens*). Both are often found in gardens and on bird-tables or feeding stations. A

Left: the Roadrunner (Geococcyx californianus) *seldom flies.*
Right: Downy Woodpecker (Dendrocopos pubescens).

somewhat similar-looking species is the Yellow-bellied Sapsucker (*Sphyrapicus varius*) which is also widespread and quite common. This species drills parallel rows of holes in trees; later it returns to these holes to devour the sap which is coming out and the insects attracted by the sap.

Other common species are the Yellow-shafted Flicker (*Colaptes auratus*) and its close western relative the Red-shafted Flicker (*Colaptes cafer*). Both are often seen on the ground where they seek out anthills and are often found far from trees. Their characteristic undulating flight and flashing white upper wings identify them at a glance. The king of the American woodpeckers is the Ivory-billed Woodpecker (*Campephilus principalis*), a very large black and white bird with a red crest. It is now only found in very small numbers in the cypress forests of the Gulf Coast. The similar Imperial Woodpecker (*Campephilus imperialis*) of Mexico is also close to extinction as its habitat of extensive stands of large trees is disappearing.

It is interesting to compare the North American avifauna with those of other continents. Whereas among the larger birds many are common to North America, Asia and Europe this is not true of the smaller birds, passerines in particular. Here many more families are shared with the South American continent and even though most of North America lies outside the tropical zones, many families are primarily tropical. In North America the mountain ranges run in a north-south direction and the West Indies and Central America form a bridge between the mainly tropical South America and the mainly temperate North America, leaving the road for invading species completely unobstructed. Thus we find many families that are only found in the New World. Such a family is the New World fly-catchers, sharing only their fly-catching habit with their completely unrelated Old World namesakes. Among the fly-catchers the Eastern Kingbird (*Tyrannus tyrannus*) is one of the most notice-

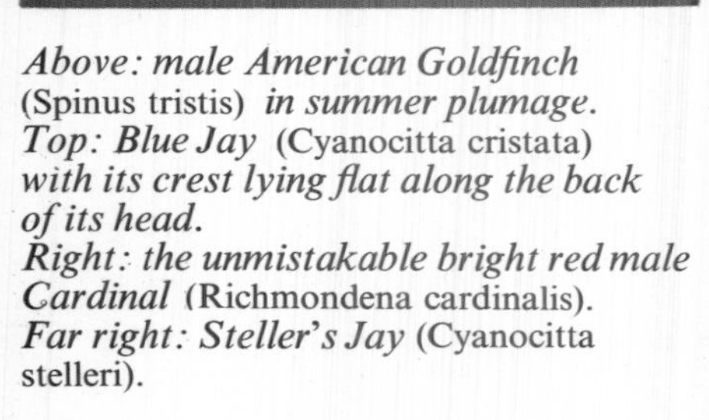

Above: male American Goldfinch (Spinus tristis) *in summer plumage.*
Top: Blue Jay (Cyanocitta cristata) *with its crest lying flat along the back of its head.*
Right: the unmistakable bright red male Cardinal (Richmondena cardinalis).
Far right: Steller's Jay (Cyanocitta stelleri).

able in the east. It is medium sized, black above, white below and is often seen chasing insects rather high in the air. It is a pugnacious bird which often attacks crows and hawks passing through its territory. Like the other kingbirds, it has a narrow red band on its crown. Its call is a loud unmelodious rasping noise. Even larger than the Eastern Kingbird is the Great Crested Flycatcher (*Myiarchus crinitus*) which is brownish with a prominently rust-coloured tail and yellow belly. It is common in the eastern woods where it nests in tree cavities. It often places recently shed snakeskins in the nest. In the south are found some brilliantly coloured flycatchers, for instance the deep red Vermilion Flycatcher (*Pyrocephalus rubinus*) and the very large Kiskadee Flycatcher (*Pitangus sulphuratus*) which has a very strongly marked head in black and white and a lemon-yellow underside. The Scissor-tailed Flycatcher (*Muscivora forficata*) is greyish-white with an extremely long streaming tail. In its aerial display the male opens and closes the forked tail rapidly. It shares the pugnacious temperament of the Eastern Kingbird. A large number of less brilliantly coloured species inhabit North America, several of them named after their characteristic calls. Two of the most common are the Eastern Phoebe (*Sayornis phoebe*) and the similar Eastern Wood Pewee (*Contopus virens*). They are best told from each other by their calls.

Only one species of the palaearctic family of larks has successfully invaded the New World, namely the Horned Lark (*Eremophila alpestris*). This species has been extremely successful in the New World and is found throughout the continent, from the tropical parts of Mexico to the Arctic. The Skylark (*Alauda arvensis*) has been successfully introduced on Vancouver Island B.C., and incidentally also on Hawaii. Another widespread family with numerous species in the Old World but poorly represented in the New are the pipits and wagtails of which only four species are breeding. The most wide-

spread of these is the Water Pipit (*Anthus spinoletta*).

Swallows are commonly seen everywhere in North America in the summer. The largest and most distinctive is the Purple Martin (*Progne subis*) which is black with a beautiful purple iridescence. Purple Martins originally nested in tree and cliff cavities, but the precolumbian Indians supplied them with nest boxes in the form of gourds. As the Purple Martin is a social nester several gourds were put up on the same pole. Today most Purple Martins nest in nest-boxes, usually a multi-compartment type on top of a pole.

The black Common Crow (*Corvus brachyrhynchos*) is found throughout most of North America. Its smaller size easily distinguishes it from the Raven (*Corvus corax*) which is also completely black. In spite of persecution by man the crow has benefited much from his activity, mainly the clearing of new land. The crows are often seen along highways, feeding both on insects found on the verges and on animals killed by traffic.

Jays of many species are found in North America. They are very colourful birds, most being blue or green. In the east the Blue Jay (*Cyanocitta cristata*) is ubiquitous. This very beautiful bird is common in gardens as well

as woods and its varied but rather unmusical call is almost an integral part of the woods, parks and suburbs. The Magpie Jay (*Calocitta formosa*) of Mexico has somewhat similar colours but has an extremely long tail and a thin but prominent crest. Recently the United States issued a stamp depicting this species as John James Audubon painted it. However, the Magpie Jay has never been encountered north of the Mexican border, and the plate was made because of a misunderstanding on the part of Audubon. In the coniferous forests of the north is found the Grey Jay (*Perisoreus canadensis*) which is of a soft grey colour. It is very tame and will visit camp-sites to obtain scraps of food. Most species of jays are found in the west and in Central America.

Tits are much less varied and colourful than in Europe. Although there are several species, these fall into two groups; one, of which the Black-capped Chickadee (*Parus atricapillus*) is characteristic, is greyish with a black cap and lip, and the other, of which the Tufted Titmouse (*Parus bicolor*) is representative, is greyish with a small crest. Nuthatches with their woodpecker-like appearance are common, the White-breasted Nuthatch (*Sitta carolinensis*) with its beautiful blue, black and white colours being the best known.

Much better represented as a family are the wrens. They vary in size and shape from the very small, chubby, short-tailed Winter Wren (*Troglodytes troglodytes*) to the rather large, much slimmer and long-tailed Cactus Wren (*Campylorhynchus brunneicapillus*) which is common in the south-western desert. Two species are found in marshes, the Short-billed (*Cistothorus platensis*) and Long-billed Marsh Wren (*Telmatodytes palustris*) filling an ecological niche taken up by a variety of warblers in the Old World.

Top left: the American Robin (Turdus migratorius) *has an orange-red breast like the European Robin, but they are not related.*
Left: White-breasted Nuthatches (Sitta carolinensis) *landing* (*top*) *and departing* (*below*).
Above: Mockingbird (Mimus polyglottos).

The Mockingbird (*Mimus polyglottos*), a medium-sized, grey, thrush-like species is the imitator among American birds. Not only does it mimic other bird calls, but it will imitate car horns, whistles and other mechanical sounds as well. It is common in gardens and sings from an exposed perch, often at night. It is frequently seen flicking its tail from side to side. It is related to the smaller and much darker Catbird (*Dumetella carolinensis*) which is also commonly found in gardens. It is named for its mewing call. It usually stays well hidden and even gives its song from the depth of a bush.

The thrashers are a group of thrush-like, very long-tailed birds with decurved bills. They spend most of their time on the ground where they find insects and snails. The Brown Thrasher (*Toxostoma rufum*) is common in the east whereas quite a number of different species are found in the west and south. Most widespread of the thrushes is the American Robin (*Turdus migratorius*), so named by homesick European settlers because of its red breast, the only thing it has in common with the much smaller European Robin (*Erithacus rubecula*). There is hardly a garden in America which does not harbour a pair of Robins or at least fall within the territory of this beautiful species. It is commonly seen on lawns as it looks for worms and its nest can be found in hedges and bushes and on ledges around houses. The juvenile Robin has a spotted breast revealing its very close relationship to other thrushes. The same is true of the spectacular Bluebird (*Sialia sialis*), of which the adult male has a blue upperside, red breast and white belly. It lives on insects caught on the ground or in the air, and inhabits farmland with trees, orchards and similar habitats. It nests in boxes or tree cavities. The Mountain Bluebird (*Sialia currucoides*) of the west is uniformly light blue. Bluebirds have found fierce competition for nesting sites in two introduced species, the Starling (*Sturnus vulgaris*) and the House Sparrow (*Passer domesticus*), both of which use boxes. Both species have been extremely successful and have spread over the entire continent where man is found.

As these two species are found wherever man is found, so the Red-eyed Vireo (*Vireo olivaceus*) is found almost everywhere where there are trees growing. This small, inconspicuous warbler-like bird is probably the most abundant bird in North America, although its favoured habitat, the deciduous forest, has disappeared in many areas. Its thrush-like song is an

integral part of spring and summer in many parts. Even more widespread, but not as numerous, is the even more indistinct Warbling Vireo (*Vireo gilvus*). Several other species of vireos are found in North America, but they are all rather plain coloured and unobtrusive. Much brighter coloured are the New World or wood warblers of which a host of different species are found as far north as bushes and trees can grow. As opposed to the Old World warblers to which they are completely unrelated, the wood warblers are limited to habitats where bushes and trees predominate. These habitats they have utilized fully, leaving little living space for species of other families. They also differ from the Old World warblers in the simplicity of their song which usually only consists of a short series of notes, with little variation. Most of the wood warblers are brightly coloured, the most common being a combination of black, yellow and white; but many species have blue and red in their plumage as well, and a few are more discretely brown like Swainson's Warbler (*Limnothlypis swainsonii*) of southern swamps. Most widespread is the Yellow Warbler (*Dendroica petechia*) which is yellow with brown streaks on

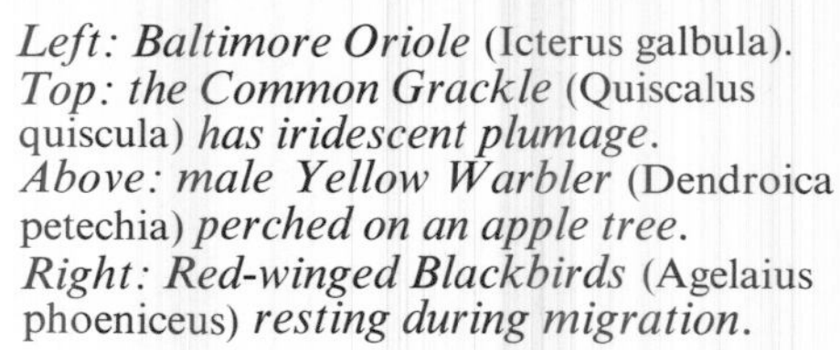

Left: Baltimore Oriole (Icterus galbula).
Top: the Common Grackle (Quiscalus quiscula) *has iridescent plumage.*
Above: male Yellow Warbler (Dendroica petechia) *perched on an apple tree.*
Right: Red-winged Blackbirds (Agelaius phoeniceus) *resting during migration.*

the breast and sides. It is common in gardens and orchards.

In autumn hundreds of thousands of warblers move south, most to South America for the winter. At this time every bush in a certain area may be swarming with warblers of different varieties, but identification is difficult as many have their dull yellowish-brown juvenile and winter plumage.

Another large family exclusively found in the New World are the Icteridae. These are medium to large, heavy-billed birds, most species being very colourful and prominent. In North America the most abundant is the Red-winged Blackbird (*Agelaius phoeniceus*) of which the male is all black except for prominent red and yellow epaulettes which he displays prominently during courtship. It is abundant in fields and marshes and usually travels in large flocks. Similar in habits is the Yellow-headed Blackbird (*Xanthocephalus xanthocephalus*) of which the male has lemon-yellow head, neck and breast and is otherwise black. It is found in the western part of the continent. Common also is the Grackle (*Quiscalus quiscula*) which is particularly abundant in farmland. It is all black with a beautiful iridescence, and has a long rounded tail. It nests colonially in dense evergreens. In the farmlands are also found Meadowlarks (*Sturnella magna*) which are short-tailed with a yellow underside interrupted by a black V. Spectacular is the Yellow-winged Cacique (*Cassiculus melanicterus*) of western Mexico; it is black with bright yellow spots on the wings, rump and outer tail and its neck feathers form a prominent crest. Like the Grackle, this bird is colonial in habits and the nest is a characteristic hanging structure. Much less spectacular in looks is the Brown-headed Cowbird (*Molothrus ater*), a common bird throughout the continent. It is rather small with a rather dull black colour. Its interest lies in its nesting habits as it is completely parasitic. Each pair has its own territory and here the female lays her eggs in the nests of sparrows, warblers and vireos. When the eggs are laid the duties are left to the foster parents, who in most cases willingly and unknowingly take upon themselves their burden. As soon as the young cowbird is hatched which, due to the

short incubation time, is before the eggs of the foster parent, it throws out the other eggs and young.

The orioles form a rather homogeneous group of the Icteridae. They are tree-dwelling, medium-sized colourful birds of about the same pattern. The males have black frontal parts, wings and tail, the rest of the plumage being bright yellow or orange. The females are much duller being of a yellowish-green colour. Most species are found in the south-west but the Baltimore Oriole (*Icterus galbula*) and the Orchard Oriole (*Icterus spurius*) are common in eastern woods where they build their hanging nests in a fork high in a shade tree. Even more colourful than the orioles are the tanagers which are somewhat similar in shape and size but belong to yet another exclusive New World family. Most, like the Scarlet Tanager (*Piranga olivacea*) of the east, are red and black but other colours are also found in members of this beautiful family; for instance the Blue-grey Tanager (*Thraupis virens*) which is common in the tropics and recently introduced in Florida.

Many of the finches found in North America are also very colourful, notably the Painted Bunting (*Passerina ciris*) which is rather common in thickets in the southern part of the continent. The male has a blue head, green back and bright red underside and rump. The widespread American Goldfinch (*Spinus tristis*) is bright yellow with a pitch black crown, wings and tail. The Indigo Bunting (*Passerina cyanea*) on the other hand is all blue. Also all blue but larger is the Blue Grosbeak (*Guiraca caerulea*), a common bird in hedgerows in the south.

Top: a Rufous-sided Towhee (Pipilo erythrophthalmus) *in flight.*
Far left: Loggerhead Shrike (Lanius ludovicianus).
Centre left: Golden-crowned Kinglet (Regulus satrapa) *taking off from a pine tree.*
Above left: the bright red Cardinal (Richmondena cardinalis) *is common in the eastern U.S.*
Left: an Evening Grosbeak (Hesperiphona vespertina) *among red maple flowers.*

Both this bird and the Indigo Bunting can often be admired as they use roadside wires for perches. Most striking is the Cardinal (*Richmondena cardinalis*). This is a medium-sized, long-tailed, all-red finch which in the east is common in gardens and parks. Adding to its charm is a small black area on the face and a rather large crest. In recent years the species has gradually spread northwards possibly aided by the many feeders of birds who in winter supplement the scarce diet of many species. There are other red finches which are less spectacular. The House Finch (*Carpodacus mexicanus*). which is abundant in the west, fills a niche somewhat similar to that of the House Sparrow. The Crossbill (*Loxia curvirostra*) is common in pine woods and in the western mountains. Above the timberline are found the beautiful rosy finches of which the Grey-crowned (*Leucosticte tephrocotis*) is the most widespread. The Rufous-sided Towhee (*Pipilo erythrophthalmus*) is a long-tailed finch which is very common in dense brush. In the west more uniformly brown-coloured towhees are also found. Most abundant are the sparrows of which there is a host of species, most of which are brown and striped. Notable exceptions are the Lark Bunting (*Calamospiza melanocorys*) which is all black with a large white patch on the wing, and the charming juncos of which the most widespread is the Slate-coloured Junco (*Junco hyemalis*). This is black with white belly and outer tail feathers. In winter its flocks are a characteristic element of wood fringes and gardens. Sparrows are found in open or bushy country, a habitat shared with the longspurs of which the Lapland Longspur or Lapland Bunting (*Calcarius lapponicus*) is the most common.

All in all North America can boast a substantial number of bird species – about 1,800 or roughly one fifth of the known species. The density of birds is also high (an estimated 20,000 million north of the Mexican border) making the continent one of the richest in the world in regard to birds.

MAMMALS

Devra Kleiman

By modern standards, the 130-million-year-old mammalian fauna of North America is surprisingly varied and abundant. With some tragic exceptions like the bison, the majority of species have resisted and survived (although with numbers greatly reduced) nearly two centuries of trapping, hunting and poisoning, as well as the continuous destruction of ecological communities caused by the industrialization of the continent. The variations in form, distribution, and behaviour among North American mammals are the result of a long evolutionary history during which species and populations have gradually adapted to surrounding environmental conditions.

The only North American representative of the primitive marsupials is the Common Opossum (*Didelphis marsupialis*), a widespread species despite its small brain and inefficient method of reproducing. Originally found only in South America, this opossum has been slowly pushing its way north, feeding on small invertebrates, lizards, snakes, cherries – in short, anything it can find. Its home is mainly the open woods and swamps where it survives by appearing only at night and 'playing possum' if caught. An uncanny ability to feign death by falling limply to the ground with its tongue drooping from one corner of its mouth has saved it from many potentially unpleasant situations.

Opossums mate in the spring, and after twelve days gestation, the female gives birth to as many as twenty bee-sized young; these look like embryos and have huge heads, forelegs with sharp hooked claws, and undeveloped hindquarters. Each must climb from the mother's vulva up to the vertical slit on the belly which leads to the pouch, its shelter for the coming three months. Unfortunately, the opossum's pouch was not built to cope with so many young, being provided with only twelve to thirteen nipples, and less than half the original number survive to

Mother Black Bear (Ursus americanus) *and her cub; not all these bears are black, some like the cub here being brown.*

maturity. Once emerged from the pouch, the opossum young take to riding on the mother's back, but they are weaned by the autumn and after this the family splits up. During the winter months individuals remain active, although they may sleep for a week or two during cold spells.

The continued survival and expanding range of the opossum is a tribute to its adaptability in the face of almost continuous persecution by man and his dogs. Another rather primitive species which is also slowly and persistently defying man by widening its range is the Nine-banded Armadillo (*Dasypus novemcinctus*). This animal is covered with a bony shell divided into a shoulder and pelvic shield at either end of nine moveable bands. An immigrant from Central America during the late nineteenth century, the armadillo has adjusted well to a North American diet of beetles, cockroaches, scorpions, worms and occasionally berries. Its strong forepaws and claws dig out juicy grubs which it can smell six inches beneath the surface, and also excavate several burrows for shelter. One of these is eventually used as a den in which the female gives birth and nurses the young. This armadillo's unique reproductive cycle begins with mating sometime in the summer months after which the egg is immediately released from the female's ovary. Once fertilized, the egg does not attach itself to the wall of the uterus, but for three to four months it remains undivided and stationary within the tract of the female. At the end of autumn, it suddenly splits into four cells which implant separately and develop into similarly sexed, genetically alike baby armadillos. From February to May, the females give birth to the quadruplets and being well developed at birth, the young soon begin to leave the safety of the burrow to follow the mother on her nocturnal jaunts through the scrub and prickly cactus thickets of their native habitat. Finally, after two months of maternal care, the family disbands.

The shrews and moles are a very widespread but primitive group of placental mammals that probably resemble the extinct mammals of the dinosaur age more than any other order. The forty or so North American species are generally small and inconspicuous, with long pointed snouts, an excellent sense of smell, and an appetite for insects. The widest ranging shrew, the Masked Shrew (*Sorex cinereus*), is rarely seen although it seems to survive in all types of habitats between Alaska and New Mexico. Feeding on insects and mice, molluscs and worms, the shrew can eat twice its body weight every twenty-four hours. Although it spends much time underground, the Masked Shrew rarely burrows or excavates its own tunnels and uses instead the pathways of other burrowing species. The female produces two or three litters a year, a total of twenty offspring. The young are helpless at birth and remain for several weeks in the nest of leaves and grasses built by the mother. Such a high birth rate is a necessity for the energetic shrew since most adults do not live longer than a year and all are vulnerable to the attacks of owls, hawks, herons, and weasels, to name but a few of the shrew's predators.

Closely related is the Northern Water Shrew (*Sorex palustris*) of the northern United States, a larger soft-furred species which inhabits damp areas near streams and swamps and feeds primarily on aquatic insects. It shows several adaptations for its partly water-bound life, among them webbed toes, fringed hind feet, and excellent camouflage while it swims; this consists of dark dorsal fur to conceal it from land predators and a light-coloured belly which looks like the water surface to underwater enemies. Water shrews use all four feet when swimming and have difficulty staying down on the stream bottom since their fur traps air bubbles which tend to make them float. On a calm day with no wind, they are said to be able to run over the surface of the water by holding globules of air in the fringe of stiff hairs around the edge of their hind feet.

The Least Shrew (*Cryptotis parva*), an insectivore weighing less than a fifth of an ounce, occurs both in dry fields and marshy regions and ranges throughout the eastern United States. It seems to be among the most sociable of the shrews since several can be caged together in captivity without fatalities; as many as five adults have been found nesting together in the wild.

The best known of the American short-tailed shrews is *Blarina brevicauda*, also called the Mole Shrew or Meadow Mole. This grey-coloured species is distributed over the eastern part of the United States. Mole Shrews move above ground and are frequently preyed upon by foxes, hawks, owls, and weasels; but they maintain their numbers by producing several litters during the summer breeding season.

Related to shrews, but larger in size and more dedicated to a subterranean existence, are the moles. The most common North American representative is the Eastern American or Common Mole (*Scalopus aquaticus*) with its naked tail and soft velvety fur. Found in open fields and pastures, woods and meadows, the Common Mole excavates its burrows at least ten inches below the soil and rarely, if ever, emerges above ground. Its sealed-over eyes, lack of outer ears, extremely sensitive snout, and cylindrical body shape are admirably suited to a tunnelling life. The mole feeds mainly on earthworms although it occasionally indulges in meals of larvae, adult insects, and vegetation. An ingenious hunter, it is known to capture its victims by covering them with soil until they either suffocate or are immobile enough to eat. Moles also kill by alternately biting at and slamming with a powerful forepaw their insect victims.

The Star-nosed Mole (*Condylura cristata*) is aptly named; with twenty-two tentacles attached to its snout, all of which are provided with sense receptors, it easily locates the worms, insects, and crustaceans that form the basis of its diet. A native of swampy pastures and low meadows in the eastern United States, the Star-nosed Mole is an excellent swimmer and is said to catch small fish on muddy stream bottoms.

Other North American moles include Townsend's Mole (*Scapanus townsendi*), the largest and handsomest mole on the continent which is able to

Left: Common Opossum (Didelphis marsupialis) *mother with her well-grown young.*
Above: Nine-banded Armadillo (Dasypus novemcinctus).

produce as many as 500 mole hills during the October to April rainy season in the Pacific states. The Hairy-tailed Mole (*Parascalops breweri*) ranges over the east coast from sea level to an altitude of 3,000 feet, and from open pasturelands to well-wooded forests and the mountainous Appalachians.

Among the most intriguing but little studied North American mammals are the insectivorous bats, adapted to fill the ecological niche left empty through an absence of nocturnal insectivorous birds. All the bats are descended from shrew-like ancestors and have developed elongated fingers joined together by a membrane which allows for true flight. Their eyesight is poor, but for navigating through the air they use a system resembling radar called echo-location. Very high-pitched sounds which humans cannot hear are emitted from the mouth (a few species use the nose), and the bat manoeuvres its way over plains and through forests by analysing the echoes that bounce back off trees and buildings. This unique method of avoiding obstacles is also employed for catching prey. As a bat swoops down on a flying insect, its cries increase in number to as many as thirty to fifty per second.

Bats living in temperate climates generally hibernate in caves or houses during the winter months since their food supply disappears at that time. Species that normally roost in exposed places, like the Red Bat (*Lasiurus borealis*) which spends its day hanging from a branch in the shade of a tree, migrate south during the autumn and hibernate where the temperature fluctuations are not so severe. The Red Bat is the most beautiful of the American species, the males sporting an orange-red coat considerably brighter than the soft chestnut fur of the female. The one or two young cling to the mother's fur with well-developed claws, and to one of her four teats with strongly recurved milk teeth. Red Bats are said to carry their young wherever they go until the combined weight of the offspring exceeds that of the mother.

The Seminole Bat (*Lasiurus seminola*) and Hoary Bat (*Lasiurus cinerus*) are close relatives of the Red Bat, the former frequenting water courses in the south-eastern United States and the latter ranging throughout temperate North America. The numerous species belonging to the genus *Myotis* are North America's most common town bats, spending the summer in colonies in attics and roofs, unlike the solitary tree bats. All hibernate during the winter in special caves where the temperature rarely varies and where the humidity is high. The best-known American species is the Little Brown Bat (*Myotis lucifugus*). Both sexes of the *Myotis* species usually hibernate together after mating in autumn, but with the arrival of spring, the females segregate themselves and establish maternity colonies in attics, roofs, churches, and in the case of the Grey Bat (*Myotis grisescens*), still in the caves. In these roosts, they bear their young in June or July. Unlike the Red Bats, the cave bats do not seem to carry their young with them on ordinary foraging trips, but deposit them instead in creches from which they are retrieved by the mother after her return to the roost.

Another of the bats hibernating in the Missouri caves with *Myotis*, but found hanging alone in drier areas of the caves, is the Big Brown Bat (*Eptesicus fuscus*). Often found in areas near man's habitations, an individual Big Brown Bat has even been sighted flying over New York City. The maternity colonies are found in houses or caves, and all the females normally give birth within a few days of each other in late June. Able to identify and retrieve its own offspring among a large cluster of squeaking juveniles, the Big Brown Bat mother will never accept an alien infant. In this respect, it differs completely from the Mexican Free-tailed Bat (*Tadarida brasiliensis*) which lives and breeds in colonies of several million in the Carlsbad Caverns in the south-western United States. In mid June, within a two-week period, one young is born to each female, and before feeding flights, the mother deposits the baby bat in an enormous nursery. Upon her return, the first two juveniles that successfully attach to the teats are accepted and cared for since she neither recognizes nor gives special treatment to her own young. Most mothers suckle two bats, but as they only give birth to one, the majority of infants are adequately reared by this system of communal maternal care.

The Lump-nosed or Big-eared Bat (*Plecotus townsendi*) is a beautiful

Left and right: Funnel-eared Bats (Natalus sp.) *hanging in a cave in Mexico. These bats have huge ears inside which are little spirally twisted false ears.*

central and western species, which is tiny and has ears reaching halfway across its body. Named for the peculiar glandular masses above its nostrils which are probably scent glands, the Lump-nosed Bat is often seen hanging suspended from cave ceilings with its ears folded against its neck.

Although the majority of North American bats are insectivorous, a number of species have become adapted to rather unusual diets, for example, the Fish-eating Bat (*Pizonyx vivesi*) and bulldog bats (*Noctilio spp.*) both of which eat fish frequently.

All twenty species of North American lagomorphs, or hares and rabbits, are strict vegetarians, subsisting mainly on grasses, twigs, leaves, stems, and berries; they inhabit every type of climate and habitat from the Arctic to the tropical rain forests of Central America. The Alaskan Pika (*Ochotona collaris*) and American Pika (*Ochotona princeps*), smaller than the rabbits and with short ears, prefer very rocky country as their home and are consequently restricted in range to Alaska, the Yukon, and the Rocky Mountains area where they excavate tortuous burrow systems beneath boulders and rocky ledges. Pikas live in large colonies, and at the approach of danger above ground, one individual will always warn the rest by uttering a high-pitched call which sends them all hurtling down into their burrows. Like the rabbits and hares, they do not hibernate, but prepare for winter food shortages by laying in a supply of thistle, gooseberry and grass. Before storing it all underground, the pika takes care to dry all the perishables under the hot autumn sun, but hoards them temporarily in rainy weather. Once dried, the grasses and twigs are

transported to a special room in the burrow system and piled up to form a haystack whose centre serves as the pika's winter sleeping quarters.

Of the ten species of North American rabbits, probably the best known is the Eastern Cottontail (*Sylvilagus floridanus*) which has a wide distribution and a range of habitats including woodlands, fields, and swamps. Despite its adaptability as a species, each individual is restricted in its movement and a cottontail's attachment to its home is so great that when chased, it would rather circle back and be caught hiding in a clump of bushes than leave its territory. The Marsh Rabbit (*Sylvilagus palustris*) and Swamp Rabbit (*Sylvilagus aquaticus*) are similar in habits and range, both species frequenting the marshy water-ridden areas of the south-eastern United States, the former inhabiting the coastal regions and the latter the borders of inland lakes and river swamps. Males are very intolerant and chase one another away from groups of females; when faced with an opponent they scratch at the ground with their forepaws. Bucks also mark territory by rubbing the secretions from the chin gland on to conspicuous objects, a behaviour known as chinning. The smallest cottontail is the Pygmy Rabbit (*Sylvilagus idahoensis*) inhabiting the arid highlands west of the Rockies where the only shelter is an occasional clump of sage-brush. With so little cover, they have taken to digging in the soft soil and are the only American rabbits to excavate and dwell in a warren, rarely being found more than a few feet from at least one burrow entrance.

Although called rabbits, the various types of jack rabbits are actually hares and are adapted for a life above ground with constant exposure to predators, a successful escape being achieved solely by the speed of their great leaps across the open prairies. The Antelope Jack Rabbit (*Lepus alleni*), which ranges through the dry deserts of the west, has slightly improved chances of escape; not only does it outdistance its enemies, but also confuses them by altering the pattern of its fur as it leaps out of sight. Ordinarily with a dark-coloured back and white sides, this hare can pull the loose white skin on its sides over its dark back and completely conceal it momentarily. This may produce a flickering difficult for a predator to focus on if the hare leaps and jumps erratically. With little water available, the Antelope Jack Rabbit feeds mainly on fleshy cactus leaves and underground tubers. The young are precocious, being born in the shade of a thicket and following the mother soon after; when only a few minutes old, they will defend themselves by rearing up on their hind legs and boxing the enemy with their forepaws.

The Black-tailed Jack Rabbit (*Lepus californicus*) and the White-tailed Jack Rabbit (*Lepus townsendi*) are both large and long-legged. The Black-tailed is the better known of the two because of its potential destructiveness. For a century, these hares have been systematically hunted by farmers who needed grazing land, and in California, whole towns used to march out beating the bushes and driving the jack rabbits (as many as 20,000 in a year) into fenced-in corrals where they were killed.

The Varying or Snowshoe Hare (*Lepus americanus*), so called because it changes from a summer coat of brown to a winter white, is distributed throughout densely wooded regions of Alaska, Canada, and the northern United States. Males and females encounter one another mainly during the breeding season when the courtship of the buck is loud and conspicuous. Rivals fight fiercely, stamping their hind feet as a threat and chasing each other through the coniferous forests.

Snowshoe Hares have long been of interest to scientists and fur trappers because of little understood cyclical changes in population which radically affect other species like the Northern Lynx (*Felis lynx*) which rely upon it for food. During population rises, does

Left: the large ears of the Black-tailed Jack Rabbit (Lepus californicus) *enable it to hear an approaching enemy from far away.*
Below left: Varying or Snowshoe Hare (Lepus americanus) *in winter pelage.*
Below: The Snowshoe Hare has very large furry feet to help it move rapidly over the snow.

produce more young per litter and more litters per year. Lynxes abound, and fur trappers become wealthy. At the end of an eight to ten year period, the Snowshoe Hares suddenly and unaccountably begin dying off after experiencing seizures, convulsions, and comas, and within a year, the lynxes are gaunt and starving. Eventually the hare population slowly builds up once again, although for many years they remain scarce.

Originally grouped together with the hares and rabbits, the rodents include the largest number of mammalian species. The lagomorphs and rodents resemble one another because they have similar styles of living, and both orders have developed chisel-like incisor teeth for coping with their primarily vegetarian diet. Among the most elegant of the sciuromorphs (squirrels and chipmunks) are the arboreal squirrels with their beautiful bushy tails functioning as balancing organs during leaps from branch to branch. Dwelling in nearly all the forested areas of North America, the tree squirrels are active during the day throughout the year and do not usually need to hibernate because they hoard food for the winter. One species which has had to adapt to the destruction of nearly all its natural habitat of hardwood forest is the Grey Squirrel (*Sciurus carolinensis*) but it has adjusted to urban life and now frequents the parks and gardens of most eastern cities. Before the cold weather arrives, grey squirrels spend many weeks hoarding acorns and nuts in preparation for the coming shortage of food. Only one nut is hidden in each small shallow pit dug by the squirrel, and these covered-over holes are distributed throughout its territory. Several months later it finds these nuts or acorns partly through its recognition of landmarks and partly through its excellent sense of smell.

The Fox Squirrel (*Sciurus niger*) occurring in the eastern United States around cypress swamps and woods, shows the same method of scatter hoarding as the Grey Squirrel, but may also hibernate for a few days during the winter if there is a cold spell. This is the largest North American species, and is rather slow-moving and heavy in the trees. The Red Squirrel (*Tamiasciurus hudsonicus*) found from Alaska and Canada south to the Appalachians, is a small attractive tree squirrel with a chestnut red coat and white underparts whose hoarding methods resemble those of the rats and mice much more than its squirrel relatives. It treats wet and dry foods differently before storing them away for the winter, always hiding cones in moist places but drying mushrooms before hoarding them. Also, the hoards are not hidden separately but are piled in one big store or larder. The closely related Western Douglas Squirrel (*Tamiasciurus douglasi*) with rusty underparts and a yellowish-fringed tail shows a similar hoarding tendency and can distinguish the good from the wormy nuts after a brief sniff, the bad ones always being rejected without even being opened.

A rare sight in the eastern and southern United States is the tiny Eastern Flying Squirrel (*Glaucomys volans*) gliding through the trees in the dead of night. It feeds chiefly on nuts and seeds, but also carrion and insects, and its gliding membrane, flat furry tail, and big eyes are a great advantage on its nocturnal food-gathering missions. During the severe winter weather it holes up in its nest but does not hibernate, preferring to feed on the acorns which it has hoarded near the nest in a hollow tree trunk. Like so many other squirrels, it threatens by stamping its hind feet and produces single chucking sounds when aggressive.

The chipmunks, favourite mammals of many North Americans, are common near urban areas and can be seen scurrying along the ground in most patches of woodland. With five dark and four light stripes along its back, the Eastern Chipmunk (*Tamias striatus*) is well camouflaged among clumps of vegetation as it pursues each meal of berries, nuts, seeds and grain. In

Left: Red Squirrel (Tamiasciurus hudsonicus) *peeping out from its leaf-lined nest in a tree trunk.*
Below left: an albino Grey Squirrel (Sciurus carolinensis). *These squirrels often inhabit parks near human habitation.*
Right: the Beaver (Castor fiber) *uses its flattened tail for swimming, and as a warning device by beating it on the surface of the water.*
Below: Chipmunk (Tamias striatus) *cleaning its whiskers.*

the autumn, the chipmunks work hard, carrying enormous quantities of acorns and nuts in their internal cheek pouches which can hold as many as seventeen hazel nuts at a time. After each collecting trip the chipmunk temporarily stores its mouthful of food in a shallow hole, but eventually all of the winter hoard is transferred to the V-shaped burrow with its one entrance. During the winter, chipmunks are rarely seen above ground, prefer-

Left: Thirteen-lined Ground Squirrel (Citellus tridecemlineatus).
Far left below: a hibernating chipmunk (Tamias sp).
Left below: Golden-mantled Ground Squirrel (Citellus lateralis).
Below and right: Flying squirrels (Glaucomys spp.) *do not fly but glide from tree to tree.*

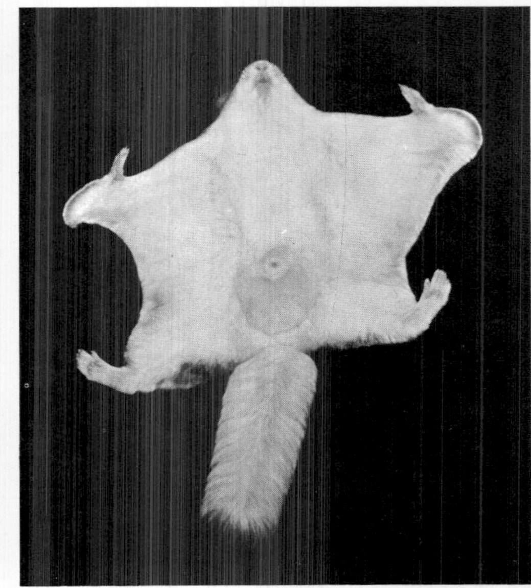

ring to rest comfortably in their warm nests with an occasional visit to their special larder for a quick meal.

Related to the chipmunks are the ground squirrels, all of which have inside cheek pouches for transporting the food which they store either in shallow surface caches or in underground storage chambers. Many of them range through the western and north-western United States and Canada, but some like the Antelope Ground Squirrel (*Citellus leucurus*) and the Mohave Ground Squirrel (*Citellus mohavensis*) are adapted for life in the Mohave desert while others like the Golden-mantled Ground Squirrel (*Citellus lateralis*) prefer the rocky mountain slopes from the Canadian Rockies south to California. All ground squirrels rely on underground burrows for shelter, probably the most complicated being the system of the California Ground Squirrel (*Citellus beecheyi*) whose tunnels may extend under fields for 140 feet with twenty separate entrances.

The Thirteen-lined Ground Squirrel (*Citellus tridecemlineatus*) although also dwelling in open fields and plains in western Canada and the United States, only excavates a simple burrow. Several animals group together to form a colony, but they seem more attracted to the habitat than to each other. Feeding mainly on grain and vegetation, grubs and grasshoppers, individuals fatten up in the summer and then hibernate, the adults going into torpor during the hot dry weather of July while the juveniles wait until autumn.

The largest group of ground-dwelling squirrels are the marmots of which there are five species in North America. Not possessing internal cheek pouches, they store food in their bodies as fat, and hibernate alone during the winter after increasing their weight by about forty per cent during the autumn. In rural areas of the eastern United States, 2nd February is known as Ground Hog Day; the Woodchuck or Ground Hog (*Marmota monax*) is said to awaken every 2nd February, returning to hibernation for six more weeks if he sees his shadow, since early sunshine means a late spring.

Extremely social are the prairie marmots or prairie dogs of the west whose enormous 'towns' or complex burrow systems once covered the plains, some being hundreds of miles in diameter and containing as many as several hundred million prairie dogs! Since the taming of the continent, the farmers have steadily reduced their numbers to provide more grazing land for cattle. The social organization of the Black-tailed Prairie Dog (*Cynomys ludovicianus*) has been studied in detail with the finding that the enormous colonies are divided up into smaller colonies whose members kiss, groom, sunbathe, and feed together, but who do not tolerate the presence of members of a foreign group. They seem to recognize each other by kissing or nuzzling in passing. The structure of each prairie dog burrow is very adaptive and provides individuals with a maximum of safety; flooding is prevented by building a mound of earth around the entrance in the shape of a lip, so that in a downpour most of the rainwater flows past and not into the burrow. In extremely wet conditions, when water actually does fill up underground chambers and stores, the prairie dog moves to a special horizontal arm of the burrow which leads to just below the ground surface, and here it can survive until the dry weather returns. The third clever addition to a burrow is a side ledge or niche which is found a few inches below the entrance; so the prairie dog can escape into the burrow but can emerge to see if danger has passed, instead of dropping down ten feet to the main underground passage.

The kangaroo rats, primarily southwestern United States desert rodents, resemble kangaroos with their long tails used as a prop and balance, long hind limbs, and short front feet. Living in arid climates, most species do not drink, but convert starch from their seed diet into water. With abundant food available only for limited periods during and after the rainy season, most species hoard large quantities of seeds in underground caches, but take care to dry the seeds before carrying them into their burrows. The Giant Kangaroo Rat (*Dipodomys ingens*) hides half-dry seeds during March in shallow pits where they are quickly dried by the heat of the sun, and then collects and removes them to a major

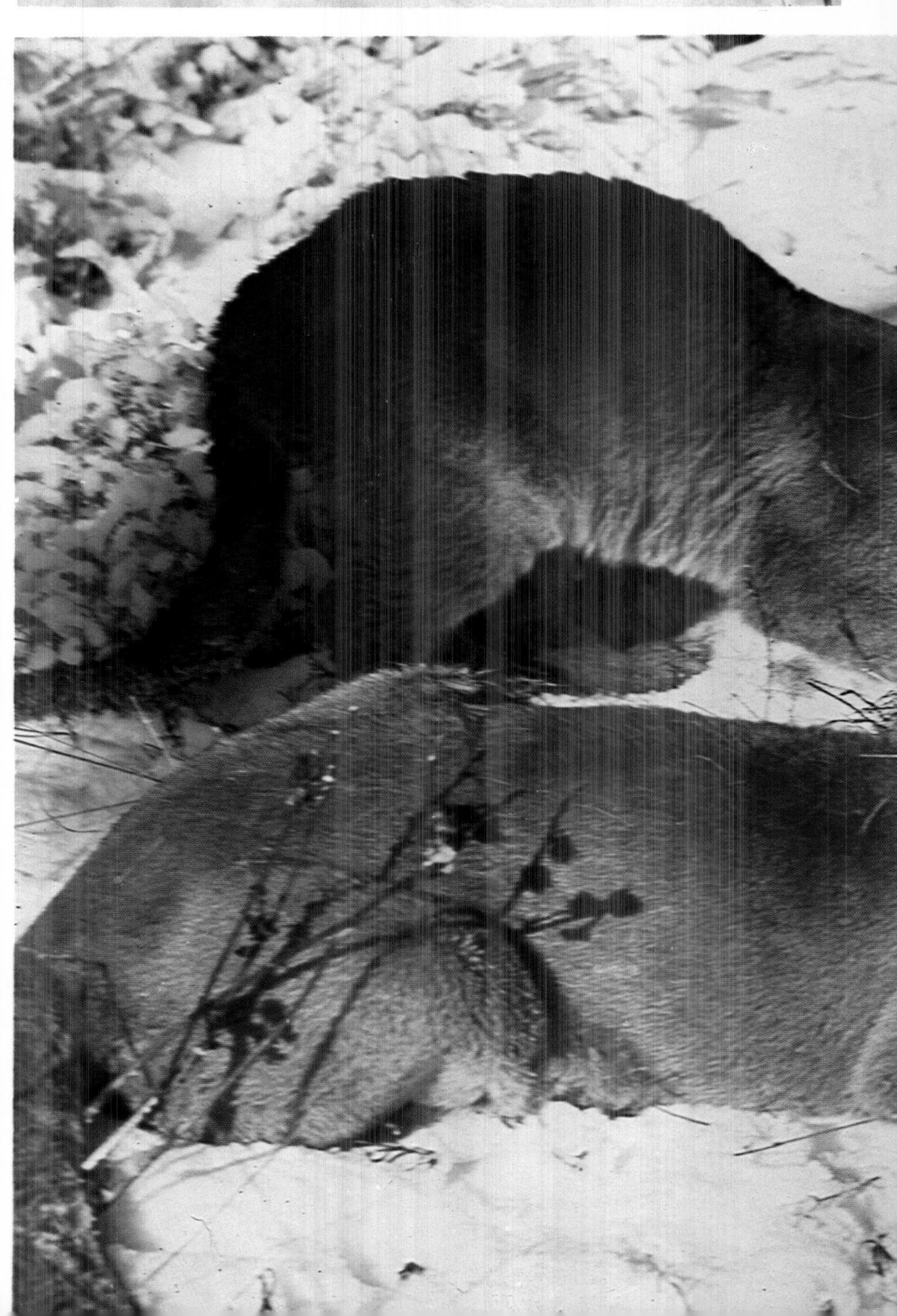

Top: Coyote (Canis latrans) *hunting in the snows of central Alaska. Bottom: female Puma* (Felis concolor) *and young, feeding on an elk.*

food store underground. Only dry seeds are found during May, and these are not temporarily stored but are immediately brought to the larder. The kangaroo rat is able to determine how much moisture is present in the seeds and deals appropriately with each type. The Banner-tailed Kangaroo Rat (*Dipodomys spectabilis*) of Mexico and the south-western United States is known for the huge mounds it builds, usually denuded of vegetation and measuring as much as nine feet by eleven feet.

One of the most primitive species of rodent is the Mountain Beaver (*Aplodontia rufa*), distributed through the west of North America from British Columbia to California. Almost tailless and with small external ears, the Mountain Beaver digs a burrow system hundreds of feet in length which contains several branches with food stores and nesting quarters. Unrelated both in structure and habits is the Beaver (*Castor fiber*), a large rodent with a flat, naked black tail and webbed hind feet. Almost exterminated during the nineteenth century, the Beaver is now considered a valuable asset for preventing soil erosion and flooding. Since the mid-1940s, Beavers have been regularly transported from valleys where they are a nuisance to farmers up to mountainous areas where they build dams in streams and rivers whose construction would have cost the government thousands of dollars. In a new area, the Beavers build both a wooden house or lodge and a dam if the water level around the home is not constant. The lodge is made of twigs and branches cemented together with mud, and has a nesting area above water level, with passages leading down into the water either to an entrance or to the food storage chamber where trees and branches are

Top: Yellow-bellied Marmots (Marmota flaviventris) *basking on a rock.*
Right: Prairie dog (Cynomys sp.) *sitting outside its burrow.*
Far right: the White-footed Mouse (Peromyscus lencopus) *inhabits woodland, feeding on seeds, plants and insects.*

hoarded for the winter. Up and down stream from the house, the dams are located, constructed of mud-cemented trees, branches, and stones. The size of a Beaver colony is restricted to an adult male, female, their yearlings, and the most recent litter; every year in the spring before the birth season, the two-year olds emigrate to build their own house and dam.

The rats and mice of North America include many species which prefer a moist habitat, but none so much as the Muskrat (*Ondatra zibethica*), a rodent resembling the Beaver with its flattened, scaly tail and webbed hind feet. Its dome-shaped houses are found in the rushes of swampy regions from Alaska and Labrador south to Mexico and are constructed of a collection of diverse water plants, rotting vegetation, and any rubbish available in the area. They are built in shallow water and the living quarters which house up to ten individuals are above water level. The Muskrat is well known for its abdominal scent glands which enlarge and secrete an especially unpleasant liquid during the breeding season.

There are hundreds of other species of rats and mice peculiar to the New World, the best known ones being the rice rats (*Oryzemys spp.*), cotton rats (*Sigmodon spp.*), grasshopper mice (*Onychomys spp.*), harvest mice (*Reithrodontomys spp.*), deer or white-footed mice (*Peromyscus spp.*), and lemmings (*Lemmus spp.* and *Dierestonyx spp.*). These species cover every possible habitat throughout the United States and Canada. Particular adaptations to different habitats can be seen in the colour of the fur of deer mice, for example; sand-dwelling mice are lighter coloured than mice living on rich dark soil deep in the forest. Primarily nocturnal, deer mice feed on seeds, grasses, and insect larvae which they hoard for the winter since they remain active and do not hibernate.

Ranging from the plains of Canada down to the Mexican border, the stocky but small grasshopper mouse is a carnivore, over ninety per cent of its diet consisting of animal foods. It mainly eats insects like grasshoppers, beetles, and crickets, but lizards and scorpions are also taken and these little mice have even been known to hunt, kill and devour field, pocket, and deer mice. They seem to live in pairs and to breed all year round. The young are not taught to hunt insects, but during a week-long period after the eyes open and before weaning, the parents tend to finish only half of their meal and leave the insect remains where the young can find them.

Less friendly with its mate is the Red Tree Mouse (*Phenacomys longicaudus*); the females live at the top of trees in the forests near the Pacific coast and only encounter a male during the breeding season when the suitor climbs up from his territory on the ground. The red-backed voles (*Clethrionomys spp.*) ranging from northern Canada to New Mexico and North Carolina are also forest dwellers, frequenting the woodlands of the north. Prolific species, they can be seen in large numbers if there has been good weather and sufficient food available.

Two species which are not indigenous to North America, but have thrived because they survive best in company with man are the Norway Rat (*Rattus norvegicus*) and the House Mouse (*Mus musculus*). Both are originally from Asia, the House Mouse being introduced about the time of the American Revolution and the rat somewhat later. The mouse has had little effect on the humans whose houses it inhabits, but the rat has caused incalculable damage to property as well as many deaths since it is the disseminator of bubonic plague, typhus, and infectious jaundice. For hundreds of years, man has been waging war against the rat by poison, trapping, and flooding its burrows, but it still thrives since man's most scientific means of control cannot cope with its ability to reproduce its kind.

Unrelated to the squirrels, rats and mice, although a rodent, is the North American Porcupine (*Erethizon dorsatum*), an inhabitant of the northern

Left: North American Otter (Lutra canadensis) *eating a fish.*
Below: Red Fox pups (Vulpes vulpes) *outside their den in central Alaska.*
Right: Black Bear cub (Ursus americanus) *sitting on a birch tree.*

woodlands. Travelling either in the trees or on the ground, the porcupine lumbers along slowly gnawing on bark and twigs during its nocturnal foraging. Although primarily a solitary species, porcupines have been found denning in family groups of half a dozen animals, and one report states that as many as nineteen have been seen sheltering together in New Hampshire. The North American Porcupine is familiar to all for its well-developed quills covering the back and tail. When confronted with an enemy, it arches its back and completely erects all the dangerous-looking spines. If the threat continues, it turns around and lashes out with its tail, which is usually enough to put off the fiercest of its predators. The female porcupine gives birth after a gestation period of over 200 days. The single offspring weighs more than a pound and is able to walk and even threaten with tail lashing within hours of birth.

Rats, mice, squirrels, shrews, and rabbits are common almost to excess in North America, but they play an important role by providing the carnivores with sufficient food; and despite persecution by man and the advance of civilization, many carnivore species have continued to flourish. Five of the seven families of carnivores are represented in the New World: the dogs, cats, raccoons, weasels and badgers, and bears.

The most strictly carnivorous of the flesh-eaters are the cats, of which there are four species although the Jaguar (*Panthera onca*) principally inhabits Central and South America and only just reaches the extreme south of the United States. The Northern Lynx (*Felis lynx*) and the Bobcat (*Felis rufa*) are quite similar to one another in appearance and habits, the Bobcat having a more southerly distribution than the lynx. Both are medium-sized spotted cats with short tails and tufts on the ears, the main differences being the Bobcat's smaller size, shorter legs,

Porcupine (Erethizon dorsatum) *mother and her single young.*

Far left: a Mule Deer fawn (Odocoileus hemionus) *has found a sleeping American Badger* (Taxidea taxus).
Left: bull American Elks (Cervus canadensis) *grazing near the Norris Geyser, Yellowstone National Park.*
Above: White-tailed Deer buck (Odocoileus virginianus).
Below: bull Moose (Alces alces) *crossing an Alaskan lake.*

longer tail, and the absence of heavily furred paws which act as snowshoes for the lynx during its travels through deep snow. Both of these cats rely for food almost exclusively upon the lemmings, hares, squirrels, voles and rats which they hunt nocturnally either by hiding and stalking or by dropping down on to the unsuspecting prey from an overhead branch. The lynx is especially dependent on the Blue, Snowshoe or Varying Hare, its numbers fluctuating as the hare population rises and falls. Shaded forests are preferred by the lynx, while the Bobcat lives anywhere there is sufficient shade and cover in mountains, swamps, or in patches of woodland. The Bobcat has shown itself to be the less specialized of the two, adapting well to the gradual replacement of its natural habitat by towns and cities.

Living in northern Canada and the Arctic region, the Northern Lynx breeds only once a year and the litter of between one and five young is born in the early spring. The Bobcat breeds in any season although only two litters are born each year. Both these cats are solitary in habit, the only social unit consisting of the mother and her offspring. Each individual has its own home range through which it wanders regularly searching for food.

Above: Jaguar kitten (Panthera onca). *Jaguars are the largest cats found in North America, although they only just penetrate the southern U.S. The young are born in late spring and reach maturity when two years old.*
Below: this Northern Lynx (Felis lynx) *has just killed a rabbit. Rabbits form the main part of its diet, but it also eats birds and snakes.*

The third and most important species of North American cat is the Puma (*Felis concolor*) which ranges from British Columbia down to the southern tip of South America. Almost leopard-sized in its northern range, the sandy-coloured unspotted Puma is capable of killing a full-grown Mule Deer (*Odocoileus hemionus*) although it also preys upon the Mountain Beaver, small rodents, skunks, and marmots. A solitary hunter, the Puma is said to take on average one deer each week and to feed from the one kill, the carcass being hidden in a cavity, until no more flesh remains. Young Pumas, unlike the adults, have spotted coats which eventually turn to a dull sandy-brown colour as the kittens mature.

The second family of carnivores with many members that are pure flesh-eaters is the mustelid or weasel group. The weasels themselves and their close relatives, the American Marten (*Martes americana*), Fisher (*Martes pennanti*), Wolverine (*Gulo gulo*) and American Mink (*Mustela vison*) are the most confirmed meat-eaters, but the kind of prey they hunt depends almost exclusively upon their size. The Least or Pygmy Weasel (*Mustela rixosa*) is the smallest living carnivore, weighing only one and a half to two ounces, and obviously unable to prey on anything larger than the smallest mice and voles, although it devours half its weight in a day. Distributed throughout eastern and mid-western United States and Canada, the Least Weasel, with its inch-long tail, appears to be more social than many of the other weasels.

The Stoat (*Mustela erminea*) or Ermine as it is called in its white winter coat, has a long black-tipped tail. Dwelling in the woodlands of northern North America and in brushland further south, it preys mainly on small mammals and birds but will

Right: the Bobcat (Felis rufa) *is a fearless hunter of rabbits, mice and other small animals. The short, bobbed appearance of the tail gives it its name.*

attack animals much larger than itself such as squirrels and hares. Within its territory a Stoat selects certain areas for carrying out particular activities, the bones of a mouse being hidden in a cavity which may differ from the nesting chamber. Captive Stoats, although extremely curious and willing to approach and sniff humans, do not like to feed while exposed and will either drag their daily rations backwards into the nest or wait until nightfall to eat. Unable to hunt on their own until mid-summer, infant Stoats are first nursed by the mother, but shortly after their eyes open at five weeks of age, the male joins in rearing them by transporting food back to the growing juveniles together with the mother.

A much larger species, the New York or Long-tailed Weasel (*Mustela frenata*) male can approach the size of a small domestic cat, although the female is only half the size and is occasionally mistaken for a separate species. Preferring to roam through open country or along the borders of marshes and swamps in the eastern United States, the New York Weasel eats small mammals almost exclusively. Its killing technique consists of piercing the area near the base of the skull with its sharp canines and severing the neck muscles. Tiny weasel cubs eat meat two weeks before their eyes open, taking care to lick their mother's lips before each meal, a special ritual of the weasel family. The cubs play intensively before leaving the comfort of the nest for a solitary adult life, many of the games resembling hunting behaviour with stalking, pouncing, and 'weasel-hugging'.

Both the American Marten and the Fisher are adapted for an arboreal life and feed mainly on birds and arboreal mammals like the Red Squirrel although the Fisher, which is as large as a fox, occasionally chooses the marten for

Opposite and above: Pumas (Felis concolor) *are also called Cougars or Mountain Lions. They live in rocky scrubby areas and are now rather rare. Right: Puma kitten.*

its victim. Both species hunt on the ground as well, and their diet can include shrews, mice and fish. The martens, leading a solitary existence, are aggressive towards one another both in and out of the breeding season although the sexes are more tolerant towards one another before mating.

The American Mink, a weasel adapted for an aquatic life with slightly webbed feet and waterproof fur, haunts the rivers and streams of the north and south-east of North America, feeding on crayfish, frogs, and fish. It mercilessly hunts the Muskrat and often moves into a Muskrat den after disposing of its inhabitants; otherwise it takes shelter under tree roots or in hollow logs. The largest and fiercest of the weasels is the Arctic Wolverine, weighing up to forty pounds. It frequents the northern pine forests west of the Rockies where it hunts mainly hares and lemmings. Wolverines are said to be able to drag a Reindeer (*Rangifer tarandus*) to the ground as well as to transport carcases three times their weight to shelter. The American Badger (*Taxidea taxus*) lives on the dry prairies from Canada to Mexico where it constructs complex thirty-foot-long burrows terminating in a grass-lined nest. Badgers emerge at dusk to hunt their daily fare of ground squirrels, field and deer mice, and gophers, with the addition of an occasional turtle egg or some insects.

Like the badger, the Striped Skunk (*Mephitis mephitis*) is nocturnal and rarely seen, but its presence pervades the fields, forests, and farmlands over which it roams. During a trip into the American countryside, the odour of angry skunk is usually overwhelming in at least one area. The foul skunk smell, a large component of which is sulphur, persists for days and arises from the secretions of the paired anal glands. When confronted with an enemy or threat, the skunk raises its tail while stamping its forefeet, turns around, jumps into a handstand, and then releases a jet of secretion which can be accurately directed as far as fifteen feet away. Fortunately, it is usually unnecessary for the skunk to complete the action since most of its predators have learned to recognize and avoid its distinctive and attractive markings. Among the most omnivorous of all mustelids, the skunk consumes insects like beetles, crickets, and grasshoppers with as much relish as birds, eggs, and small rodents. During autumn, fruit is added to the diet, and the skunk becomes overweight in preparation for the winter when it sleeps much of the time although it does not enter true hibernation.

The North American mustelid most adapted for an aquatic existence is the North American Otter (*Lutra canaden-*

Above right: Long-tailed Weasel (Mustela frenata) *in its winter pelage; in summer it is reddish-brown with white underparts.*
Right: American Badger (Taxidea taxus), *a powerful animal that can defend itself against almost all carnivores.*
Far right top: a litter of baby Striped Skunks (Mephitis mephitis).
Far right below: Striped Skunk. The raised tail and erected back fur are warning signals that it is about to squirt its foul-smelling secretion.

sis) which is found north of Mexico anywhere there are lakes and large rivers. It is distinguished by its short hind legs and webbed hind feet, cylindrical body shape, and strong tail. Although normally not a wanderer since its diet of trout, crayfish, frogs, and clams can be found close to home, the otter must take prolonged journeys overland when its normal food supply is trapped under frozen-over waterways during severe winter weather. When moving across ice, the otter is said to slide instead of walk; adults and young also spend considerable leisure time sliding down steep banks into rivers and lakes, carefully removing every shred of vegetation from the bank to make it more slippery. Otter tobogganing, a social sport since individuals have never been seen sliding alone, appears to be without any serious function.

During April and May, the helpless young are born in a burrow near the river bank, sometimes beneath the thick roots of an old tree, and the mother cares for them on land until they are near weaning age. Not natural-born swimmers, otters must be persuaded to enter the water by the mother and encouraged to swim, but their swimming technique is soon indistinguishable from the adults'. Otter young must also be taught to feed, and the mother catches prey live and releases it in front of the juveniles after calling them to her with a special feeding call; she then stands by watching as they improve their killing abilities.

The raccoon family is exclusive to the New World, except for the panda members. Its best known member is the North American Raccoon (*Procyon lotor*) with its black-masked face and long black-ringed bushy tail. An inhabitant mainly of the forests, and frequently found hidden in trees, the raccoon can also become adapted to life on sand dunes, in deserts, or in swamps. It feeds on anything from crayfish to corn and fruit, and in captivity has always been considered a fastidious feeder since it often washes its food before eating. It has recently been found that this unusual behaviour is not related to cleanliness at all, but is simply an expression of the raccoon's desire to search for shellfish on a sandy, gravelly stream bottom using its sensitive forepaws and fingers to distinguish between the rocks and clams.

Relatives of the raccoons, although looking quite distinct, are the bears of which there are three species in North America, the American Black Bear (*Ursus americanus*), the Brown Bear (*Ursus arctos*), and the Polar Bear (*Thalarctos maritimus*). All are large dangerous beasts which have given up a strictly carnivorous way of life (only the Polar Bear still eats primarily flesh) but are more dangerous than any other meat-eater because they are unpredictable and will attack man without warning if disturbed. The absence of a threat like a growl or screech is related to the fact that most bears are completely solitary and only rarely encounter one another over their wide ranges; they therefore have no need for an aggressive vocalization or gesture which will intimidate a rival.

The Black Bear, the smallest and commonest of the American species, dwells in the unsettled forested areas from the New England Adirondacks to the Rockies. Feeding on a variety of foods, including nuts, fruits, bark, small mammals, and bees' nests, the Black Bear spends the autumn months gorging itself in preparation for its long winter sleep during which it does not hibernate but simply remains torpid. Black Bears are familiar to American tourists who see them begging for food in large national parks like the Yellowstone.

One race of Brown Bear, the Grizzly, is probably the largest living land carnivore (it weighs up to 1,500 pounds) and although near extinction,

it is still found in some uncivilized areas of the west. A nocturnal hunter, the Brown Bear feeds on salmon in nearby streams and also attacks sheep and other large mammals, killing them more by sheer power than technique. Brown Bears are solitary and roam over set pathways in their territories which cover more than ten square miles. Both visual and scent marks are distributed on trees within the bear's home range, leaving signs of the individual's presence which may intimidate rivals or intruders. Mating occurs in mid-summer, but implantation of the fertilized egg is delayed and the twin cubs are not born until mid-winter. The birth of the cubs, which weigh only one pound each, occurs while the mother is torpid, but by the time she emerges from the winter den the young are ready to follow her, although they do not reach maturity for several years.

The breeding pattern of the Polar Bears, which are distributed around the Arctic region, are similar to that of the Brown Bears, one or two tiny cubs being born in mid-winter. The young develop slowly and remain with their mother for up to two years, by which time she is ready to have another litter. Polar Bears are flesh-eaters preying primarily upon seals which they stalk and attack from ice floes, beating the victims to death with their forepaws. Polar Bears wander far with no preferred home base, and they cover long distances over both land and sea,

Far left: the webbed feet, streamlined body and powerful tail of the North American Otter (Lutra canadensis) *are all aids to swimming for this largely aquatic animal.*
Below: the Sea Otter (Enhydra lutris), *which was once hunted nearly to extinction for its valuable pelt, is now protected and is gradually increasing.*

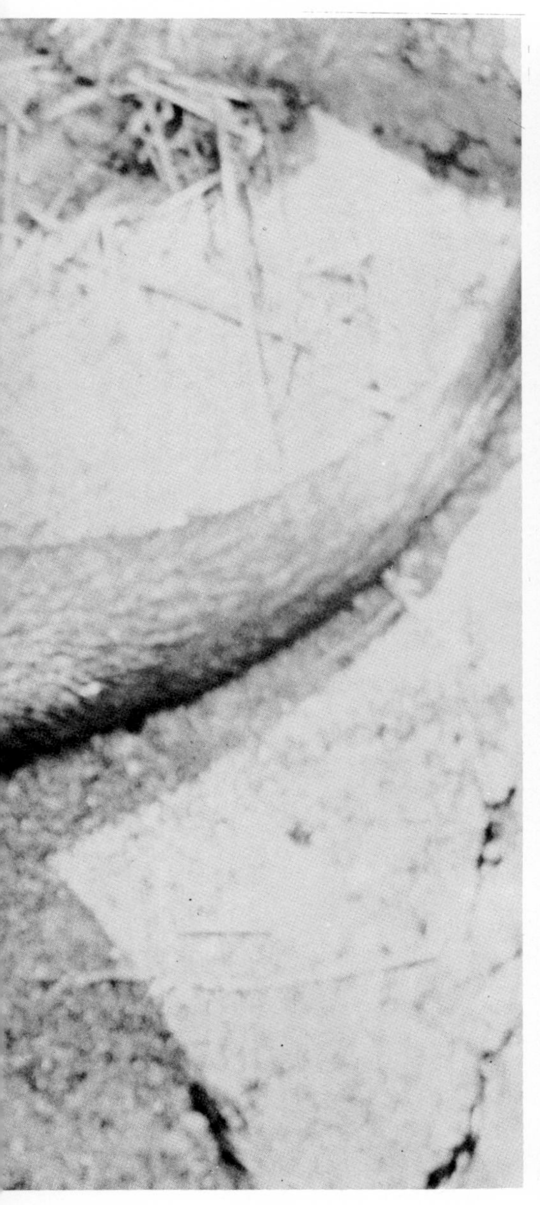

the latter by swimming for which the Polar Bear is well adapted.

Among the most non-specialized and therefore adaptable of all carnivores are the canids (dogs, wolves, and foxes) with nine different species present in North America. The most wide-ranging is the Red Fox (*Vulpes vulpes*) with its long black-tipped bushy tail, large pointed ears, and slim body. A mainly solitary creature, the fox catches its prey by a combination of stalking and chasing depending upon the size and speed of the victim. Usually field and deer mice, rabbits, and hares are hunted, but when there is a shortage of small mammals, the fox quite readily turns to eating insects, berries, roots, and other vegetation. The numerous fox vocalizations consisting of growls, several types of barks, whimpers, and squeaks, are extremely important during the breeding season since they all differ in meaning and are used to communicate the mood and reproductive state of both dogs and vixens. After mating in January or February, the dog fox remains with the female during the gestation period and eventually helps

Right: Black Bear cub (Ursus americanus) *climbing a tree to escape danger; cubs are agile climbers but adults rarely ascend trees.*
Below: North American Raccoons (Procyon lotor) *wading a stream.*

to wean the offspring by carrying food to the female and young in the middle of summer.

More sociable than the Red Fox is the Arctic Fox (*Alopex lagopus*) which ranges across the Arctic and relies on the lemming populations for food. The only canid with a seasonal change in the colour of its fur, the Arctic Fox is bluish-grey during the summer and a dazzling white in winter. In its snow-covered home, this species is said to be colonial with numerous interlocking burrow systems shared by several animals. Captive individuals are more tolerant of each other than Red Foxes, and two females have shared a den and jointly reared their two litters with no aggressive interactions. Not a great deal is known about the behaviour of the two species of grey fox (*Urocyon spp.*) and the Swift or Kit Fox (*Vulpes velox*).

Completely different in their habits are the dogs, especially the Wolf (*Canis lupus*) whose shrinking numbers and ever-decreasing range throughout northern Canada and Alaska has been caused by a lack of sufficient food and by persecution from man. Originally found throughout the plains country of North America, the Wolf was steadily pushed north as the immense bison herds were wiped out during the nineteenth century. Unlike many smaller carnivores, it could not adapt its specialized pack hunting techniques and dependence on large game such as deer to conditions of civilization.

Wolves are extremely social creatures, preferring to live in packs of six to twenty-five individuals. Each group has a hierarchy or pecking order which is long lived and accepted by most members, so that aggression or disagreements arise very rarely. The general friendliness is expressed in many ways, but is best seen before the evening hunt when the whole Wolf

Top: young North American Raccoon (Procyon lotor) *feeding.*
Left: Black Bear cubs (Ursus americanus) *are usually born late in the winter and are ready to leave the den in late spring. This is the most common American bear.*

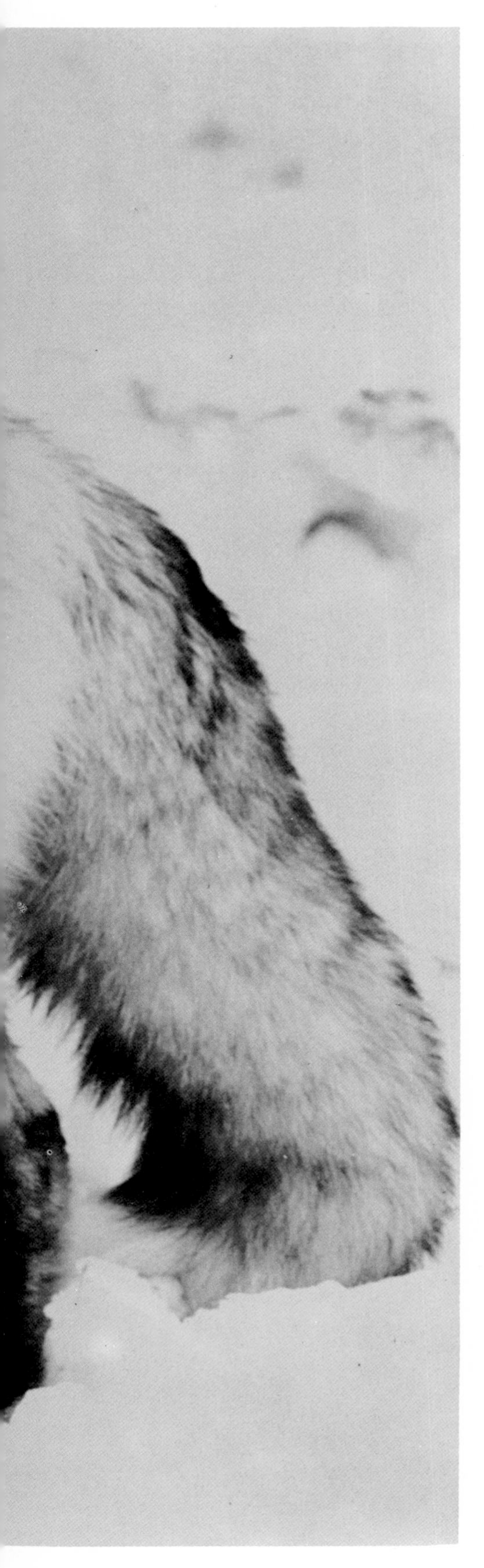

pack will come together in great excitement, tail wagging, sniffing and licking each other's lips and muzzles, and often howling. Different postures and movements of the ears, mouth, body and tail are used to great advantage to express changes in moods which helps to avoid misunderstanding between individuals, a necessary development in a social species as powerful and potentially dangerous to itself as the Wolf. Attacks are never unpredictable as in the solitary bears, but are preceded by threats, usually teeth baring (or snarling) and growling which possess subtle differences representing variations in the strength of the aggressive tendency. Individual Wolves can more or less measure the degree of aggression present in a threat and will respond in the most appropriate way by fleeing, behaving submissively, or challenging their attacker.

Before the breeding season in late winter, the social interactions among pack members increase, with more frequent howling and occasional challenges directed towards the most dominant male. Pairs may form or be reformed between individual males and females, but many young females may eventually mate with several males. The young are born in a den dug by the pregnant female, and if the pack has remained together, several members will take turns staying with the young if the mother decides to join the hunt one night. Both adult males and unmated females hide meat in the vicinity of the den, and when the cubs are near weaning age, several pack members may join in feeding the juveniles partly digested flesh which is regurgitated.

The Coyote (*Canis latrans*) resembles the Wolf in many ways although it is smaller and more solitary in habit. During the past fifty years, it has slowly taken over much of the range once occupied by the Wolf; whereas it was once only found on grasslands and prairies in the west, it now stretches across the continent from the east to the west coast. Its success is based on its unspecialized food habits since,

Left: Red Fox (Vulpes vulpes) *with its prey, a cottontail rabbit* (Sylvilagus sp.). *It also feeds on small mammals.*
Above: the Wolf (Canis lupus) *is now only abundant in Canada, having been exterminated from most of its former range which extended into Mexico.*
Below: a young Wolf pup.

Above: Walruses (Odobenus rosmarus) *rushing into the sea in fright.*
Right: Steller's Sealion (Eumetopias jubatus) *bulls are very large; this one may weigh 2,000 lbs, the pup 50 lbs.*
Centre right: the Rough-toothed Dolphin (Steno bredanensis) *is found in temperate waters throughout the world.*
Far right: mother and baby Bottlenosed Dolphins (Tursiops truncatus). *These intelligent mammals are often kept in large marine aquaria where their playfulness delights the public.*

like the foxes, it feeds on a variety of small mammals as well as vegetation and insects.

Related to the carnivores but adapted for an aquatic life are the seals and sealions, known as the pinnipeds and found along both the Atlantic and Pacific coasts of North America. The body shape of the seals has been radically modified for a waterbound life, limbs have been replaced by flippers for efficient propulsion through the water, the tail has been reduced or

has disappeared, and a thick layer of fat under the skin serves for insulation instead of fur and also provides energy during the female's lactation and the male's self-induced summer starvation. The seals subsist primarily upon non-commercial fish, molluscs, and crustaceans, but have sometimes been accused of and killed for feeding on salmon. In turn, they are preyed upon by Polar Bears, sharks, the Killer Whale (*Orcinus orca*), and especially man, who has brought several species close to extinction. These aquatic mammals have extensive ranges, most species migrating north every year for the breeding season and then south again for the winter. The Northern Fur Seal (*Callorhinus ursinus*) with its blunt bear-like face travels as much as 3,000 miles through the Pacific to reach its breeding grounds in the Bering Sea Islands.

The seals are best known for their unusual breeding habits. In the majority of species, huge numbers congregate on the beaches of small islands and coastlines, most individuals returning to the breeding grounds of their birth. Territories are established and maintained by the males threatening and fighting, and once the females arrive harems are formed from the number of cows a male can guard and defend. The number depends on the size of his territory which is partly regulated by his bulk. Northern Elephant Seal (*Mirounga angustirostris*) bulls, with an unwieldy weight of 5,000 pounds, move poorly on land, and this restricts the number of cows in their harems to about twelve. Steller's Sealion (*Eumetopias jubatus*) males, whose weight ranges up to 2,200 pounds, can guard slightly more females, but cannot compete with the California Sealion (*Zalophus californianus*) and the Northern Fur Seal which, weighing only 500 pounds, can command as many as forty females at a time. Even with the harems established the males continuously defend their females and territories from rivals, with the result that they do not feed for two months during the breeding season, living off their stores of blubber.

Soon after arrival at the beaches, the females give birth to one offspring; reproduction is so well-timed that in

Above: Musk Oxen (Ovibos moschatus) *live in the tundra of northern Canada where they are hunted by the Eskimos.*
Right: a herd of American Bison or Buffalo (Bison bison); *now only small groups remain of the huge herds that once roamed the prairies.*
Below: male American Bison.
Far right top: White-tailed Deer bucks (Odocoileus virginianus) *sparring.*
Below right: White-tailed Deer with antlers still in velvet.

the Northern Fur Seal a pup is born every five seconds during mid-July. Mothers, recognizing and responding to their own pups among thousands of similar-looking young seals, usually suckle for many months. The Harp Seal (*Phoca groenlandica*) mother, however, is said to abandon her offspring after only two weeks of nursing, and after a fortnight of fasting, the youngster sheds its white baby fur and begins feeding on the crustaceans and fish that form the bulk of the adult's diet.

As the summer ends, the females and pups, abandoning the males and beaches, begin their long migration southwards for winter.

The big North American carnivores like the Wolf and Puma developed as predators for the once enormous populations of ungulates, including deer, American Bison (*Bison bison*), Musk Oxen (*Ovibos moschatus*), and Pronghorns (*Antilocapra americana*). All of these species are even-toed ungulates, and have a complex stomach with several chambers allowing them to feed quickly and digest slowly by 'chewing the cud' or ruminating. They have horny outgrowths from the top of the skull of which there are several types, the deer having seasonally occurring branched antlers which are skin-covered as they grow, and the bovids (cattle, sheep, and goats) having permanent horny growths which are unforked and usually found in both sexes. The Pronghorn, the sole representative of a very ancient family of North American ungulates, sheds its horns annually like the deer but maintains a permanent bony core. Not true antelopes, the buff-coloured Pronghorns live in small herds in dry rocky desert country and feed on sagebrush and cactus. Bucks collect up to twenty does in August before rutting, and eight months later, the twin kids are born. Females deposit their twins in separate hiding places during the first few days, but soon the offspring begin to follow the mother. The Pronghorn is the fastest North American mammal, defending itself from Coyote and Wolf predation by outrunning its attackers.

The Black-tailed or Mule Deer (*Odocoileus hemionus*) and the White-tailed Deer (*Odocoileus virginianus*) range throughout North America although the former is more restricted to the partly wooded habitats of the west and avoids deep forest. In both species, males and females browse together in large groups during the winter months, feeding on any available vegetation. As spring approaches, the groups disband and females give birth in early summer to twin fawns whose spots camouflage them as they lie hidden during the day while the mother forages for food. Throughout the summer months when the antlers are growing, the males are solitary or living in small groups of bucks.

The Reindeer or Caribou (*Rangifer tarandus*) is the most northerly species of deer distributed throughout Alaska and Canada, and both males and females possess branched antlers. A migratory species, the Barren Ground Caribou (*Rangifer tarandus groenlandicus*) covers hundreds of miles while travelling from its winter to its summer quarters in the rich tundra, where the does give birth and there are rich supplies of lichen, moss, and shrubs for food. Although there were more than two million in 1900, Caribou numbers have been drastically reduced with only several hundred thousand surviving today. Rutting occurs during the southerly autumn migrations, and the bucks shed their antlers soon after; females, however, maintain their antlers until March thus keeping their status higher than the males and ensuring that they get first choice of food during the winter while they are pregnant. Both very young and old and weak Caribou are persistently attacked by packs of Wolves which follow the great herds on their migrations. Rarer than the Barren Ground Caribou is

the Woodland Caribou (*Rangifer tarandus caribou*) which roams through the forests of northern Canada, coming out from the woods to wade in rivers and lakes during hot summer days.

The largest living deer is the Moose (*Alces alces*) with its long ungainly legs, fleshy proboscis, and spoon-shaped antlers (worn only by the males). A browser rather than a grazer, the Moose wanders in small groups feeding on bushes and shrubs during the summer. Males are aggressive during the autumn rut, making fearful bellowing noises and crashing through the trees as though ready to attack any living creature that approaches them. Males fight by charging each other with the head lowered. Some have been found dead with their antlers locked together, obviously the result of a battle where the two males could not disengage from each other. During May, Moose calves are born, and the mother becomes very aggressive, kicking with her powerful front legs at anything that disturbs her.

Left: Pronghorn Antelopes (Antilocapra americana).
Below: bull Moose (Alces alces). *Moose live in damp areas of Canada and the northern United States, often feeding on water plants.*
Bottom left: Pronghorns are only found in the western plains of North America and have no close relatives.
Bottom right: cow Moose.

The Wapiti or American Elk (*Cervus canadensis*) ranging through the western United States and Canada, differs from the other deer species in its social organization; the males are more polygamous, strong bulls collecting sixty or more cows for their harems during the autumn rut. These females are diligently guarded from the advances of other males, and before attacking an intruder, a bull will signpost his territory by scraping the bark off a tree with a burr at the base of the antler and then rubbing his nose up and down over the scraped area. During the winter, the elks travel in large herds from the highlands down into the valleys in search of food.

The American Bison, now only found on reserves, once covered the prairies from Canada to Mexico in huge herds, but was wiped out during the nineteenth century when hunters like Buffalo Bill Cody were between them killing thousands per day. Its extermination seriously affected the livelihood of most tribes of American Indian, who had relied upon the bison for food, clothing, shelter (hides were used for making tepees), food implements, ropes, and glue. While shedding its dark heavy winter coat, the bison wallows in prairie dust, rubs against trees, and plasters itself with mud as protection against insects.

The Rocky Mountain Goat (*Oreamnos americanus*) also rubs mud all over its body, but mainly as part of rutting rituals. Normally, billies defend several females against rivals by lateral fighting where each individual attempts to strike at its opponent's forequarters and chest. Actual blows are rare, but in a really serious battle, one billy can disembowl his rival. The Rocky Mountain Goat roams in small herds through the steep mountains and cliffs of western Canada, subsisting on shrubs, lichen and moss and rarely descending into the valleys. Two other North American ungulates are the Musk Ox (*Ovibos moschatus*) an inhabitant of the Canadian tundra which resembles the central Asian Yak (*Bos grunniens*) with its heavy coat of hair; and the Bighorn Sheep (*Ovis canadensis*) distributed from western Canada to Mexico where it frequents rocky regions with sparse vegetation.

Left: White or Dall Sheep (Ovis dalli) *are found in the mountains of western Canada and Alaska. They are very sure-footed, being able to climb steep narrow slopes and ledges.*
Above: Bighorn Sheep (Ovis canadensis) *are found in a few areas of the Rocky Mountains but were once much more common.*
Right: a herd of Barren Ground Caribou (Rangifer tarandus groenlandicus) *on a summer snow patch to escape flies.*

REPTILES Brendan Kypta

The vast reptile population of the North American continent provides a veritable herpetologist's paradise. Great care is being exercised by the various public bodies, through passing laws and fostering a great public interest, in preventing the depletion of America's animal heritage. The attempts at conservation of the 'herptiles' (a group name given to reptiles and amphibia) are largely carried out through the dedication of the huge number of active herpetologists, both amateur and professional, who research into all phases of the study of these interesting animals. In well equipped laboratories, in mosquito-infested swamps, and in oven-like deserts this work goes on, and it is generally accepted that more is known about the behaviour, courtship and breeding, and longevity of the American reptiles than of any other group in the world. Many scientists have devoted their whole life to the study of a particular family, an outstanding example being Dr. Laurence Klauber's 'Life Histories and Habits of Rattlesnakes'.

The rattlesnakes are a unique group of snakes belonging to the New World, and are found from Canada to Panama. Only one species, the Cascabel (*Crotalus durissus*), extends into South America. This fact is surprising as there are thirty species living in North America. Some grow to more than eight feet long, such as the Western Diamondback (*Crotalus atrox*) and the Eastern Diamondback (*Crotalus adamanteus*), but the Pygmy Rattler (*Sistrurus miliarius*) and its sub-species rarely grow longer than fifteen inches. The rattlesnakes derive their name from the appendage on their tail. At birth it resembles a small button, but when the snake sheds its skin a further horn-like segment forms, and this in time produces a series of horny rings which create a rattling or buzzing sound when the tail is vibrated. Large specimens can be heard some distance away, but the smaller Pigmy Rattlers sound more like the buzzing of a bee.

The Canebrake Rattlesnake (Crotalus horridus atricaudatus) *lives in mountain forests and the eastern coastal plain.*

The rattlesnakes are venomous and are responsible for about forty-five deaths each year out of about 1,500 bites from various poisonous snakes. A large diamondback rattler suddenly confronted in the open can become a terrifying spectacle. It coils in neat circular formation with head and tail raised from the centre, the tail vibrating and the head poised ready to strike, like a tensed spring. The tongue flickers in and out of the mouth through a small aperture between the jaws, with the rattling intermittently hesitating and then increasing in tempo. If left alone, the head sinks down, the rattling ceases, and in a while the snake will glide down a hole in the ground or under a rocky shelf. Many of these snakes take up residence in the burrows of marmots and gophers, and many sentimental stories have been told of friendship between the rightful owners of the burrows and the snakes; but examination of the stomach contents of the snakes has disproved these stories – the owners' offspring were on the menu! Occasionally it has become necessary to reduce the numbers of these poisonous snakes in some districts where they have increased too numerously. In some States, yearly hunts take place to capture large numbers as they emerge from their hibernating chambers under rocks. In at least one State, dynamite has been used to destroy excessive numbers in inaccessible caves; a captured snake was used to carry unwittingly the charge and fuse into the cave where its hibernating companions were coiled up. This drastic method of control is usually resorted to only when others fail, but large numbers of these very dangerous snakes in the vicinity of an otherwise pleasant camping site is bound to call for some effective remedy.

The rattlers are attractively patterned snakes, and although not brilliantly coloured except in the case of the Pigmy Rattler of Everglades (*Sistrurus miliarius*) which sometimes has a red

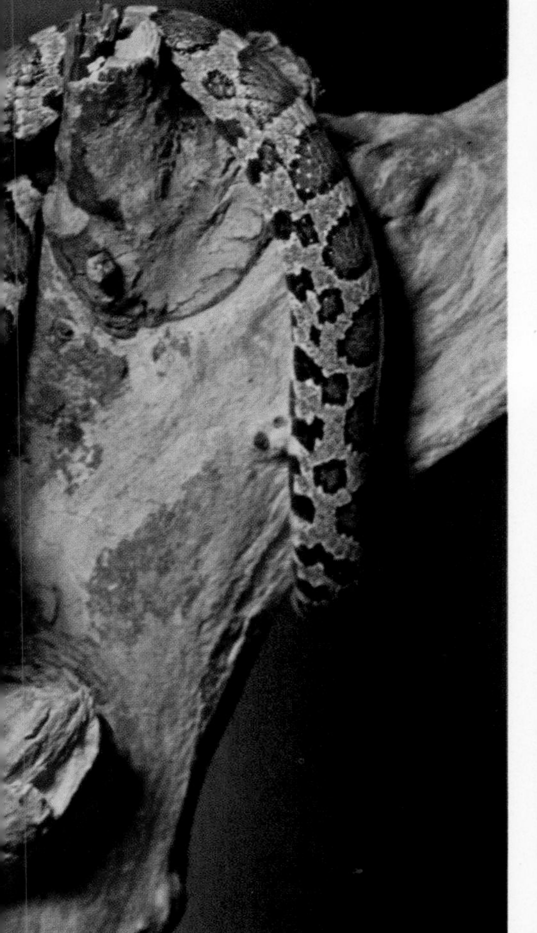

Far left top: a tightly coiled Canebrake Rattlesnake (Crotalus horridus atricaudatus).
Far left bottom: the Copperhead (Agkistrodon contortrix) *is one of the best known poisonous snakes.*
Centre: the Red King or Milk Snake (Lampropeltis triangulum).
Above: Speckled King Snake (Lampropeltis getulus holbrooki) *of the Mississippi Valley.*
Left: the harmless Scarlet King Snake (Lampropeltis doliata) *gains protection from the similarity of its colouring to that of the poisonous coral snakes.*
Below: the Corn Snake (Elaphe guttata) *is often found in corn fields.*

stripe, all have rich velvety colours of browns, creams and yellows.

The Timber Rattler (*Crotalus horridus horridus*), a yellowish snake with brown bands or blotches across the back, was exceptionally numerous in the eastern States, but has been greatly reduced in numbers except in more remote mountain districts.

The closely related Canebrake Rattler (*Crotalus horridus atricaudatus*) is very similar to the previously mentioned snake, but can be identified by a chestnut brown stripe down the centre of the back, from the head to about halfway down the body. The Eastern Diamondback and the Western Diamondback are the largest of the group and the most dangerous of the North American snakes. The Eastern lives in Carolina and Florida near the coast or in the wooded areas, and is sometimes observed swimming in the sea.

The Western is the rattler of the desert and canyons of California, Arizona and Texas, and south as far as Mexico. The diamond marking on its back is not so pronounced as in the former snake, but on my few encounters with it I have always found it more aggressive than any other rattlers. It rarely retreats but goes on attacking to the last.

The Prairie Rattler (*Crotalus viridis*) is a more tractable snake and will generally glide away if given the choice. As its name implies, it is the snake of the great plains, but it has a widespread distribution from Canada to Mexico.

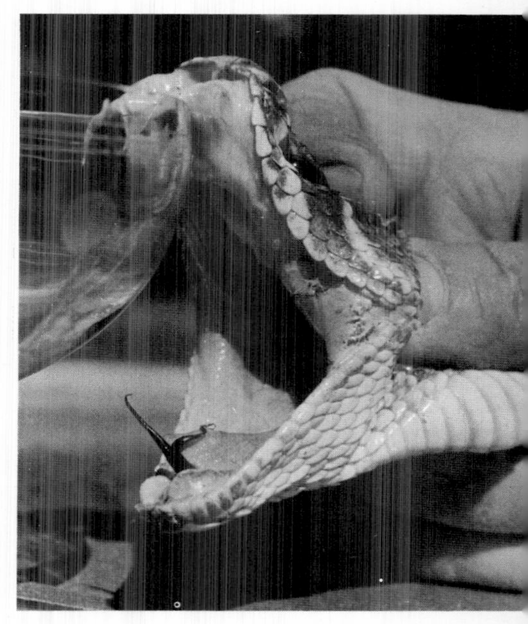

Above left: newly hatched young Timber Rattlesnakes (Crotalus horridus).
Left: milking a Diamondback Rattlesnake (Crotalus sp.) *for the preparation of serum against bites.*
Above: these two male Red Diamondback Rattlesnakes (Crotalus ruber) *are engaged in a combat display dance. It soon becomes obvious which of the two is the aggressor.*

Its body colour is greenish-yellow, and down the centre of the back round blotches of brown are widely spaced, with smaller brown spots along the sides.

In outlying districts fallacies relating to snakes are still repeated; probably these are derived from Indian folklore in which some snakes were revered while others were feared. One story that fairly persistently crops up is the 'hoop snake' story in which these imaginary serpents place their tail in their mouth to form their body into a circle and roll like a hoop at high speed after their victims. The story may have originated when a Sidewinder was first seen..

The Sidewinder or Horned Rattlesnake (*Crotalus cerastes*) deserves a special mention. This pale sand-coloured snake which lives in the deserts of Arizona and parts of California and Nevada, has developed a method of locomotion unlike that of any other American snakes. When it travels slowly it moves as other snakes do in a more or less straight line, but when in a hurry it throws its body forward in horizontal loops, which enables it to travel at a surprising speed over loose sand. It is strange to find that the only other snakes that move in this way are the Cerastes Viper (*Cerastes cerastes*) and a few near relatives which live in the deserts of North Africa and the Near East countries.

The rattlesnakes are live-bearing snakes giving birth to a dozen or more young which are usually brighter in colour than the parents. They are equipped with poison fangs and a supply of venom at birth. The effectiveness of their venom was proved some years ago when one of a consignment of these snakes unexpectedly produced a family during transit to a British zoo. During the unpacking an attendant was bitten by a baby rattlesnake a few

Top left and bottom left: Garter snakes, swallowing a dace and floating on a mat of algae.
Top centre: Red-eared Elegant Sliders (Pseudemys scripta elegans) *sunning themselves.*
Top right: adult and young Short-horned Horned Toads (Phrynosoma douglassi); *the two smallest have not yet developed their spines.*
Bottom right: a 2-foot-long Eastern Coral Snake (Micrurus fulvius), *a small but very dangerous species.*

days old, and spent several months in hospital as a result.

There is another small group of pit vipers in North America in addition to the rattlesnakes. These are the venomous Cottonmouths and Copperheads which comprise the five subspecies that make up North America's representation of the genus *Agkistrodon.*

The largest of these are the Eastern and Western Cottonmouths (*Agkistrodon piscivorous piscivorous* and *Agkistrodon piscivorous leucostoma*). These impressive and robust vipers, which can grow to a length of more than five feet, are mainly aquatic. They have large heads which are carried high and often the mouth is held open giving them a frightening appearance. Old adult specimens are dull brown or nearly black, but the young are brighter chestnut brown with darker brown bands. At all times they blend into their surroundings and are often unnoticeable as they lie on some old rotten tree stump or moss-grown rock in the swamps where they live. They also vibrate their tails when annoyed, but since the tail has no rattle, little noise is made unless it is in contact with some dry vegetation and then it can be mistaken for one of the smaller rattlers' warnings.

The Cottonmouths are omnivorous and whereas the rattlers are chiefly mammal and lizard feeders, the Cottonmouths will eat other snakes, fish, mammals, birds and frogs. I have known the very young to eat even large fleshy insects.

The Copperheads (*Agkistrodon contortrix*), mainly reddish-brown snakes with dark brown hour-glass markings along the back, are found in a great variety of habitats from Massachusetts to the Gulf States. Sometimes they may be seen in marshy swamplands, but in general they prefer woodlands and rocky parts where the small

Top: this Eastern Diamondback Rattlesnake (Crotalus adamanteus) *is about to shed its skin.*
Right: Prairie Rattlesnake (Crotalus viridis) *shedding its skin.*

rodents upon which they feed can be found, though the individuals living in damper areas appear to feed mainly on frogs. Copperheads and Cottonmouths are both viviparous, usually having small families of six or eight and never as many as some of the rattlesnakes which occasionally have large litters of up to twenty.

North America possesses other venomous snakes belonging to the elapine family of snakes, of which the deadly cobras of Africa and Asia are members; these are the Common Coral Snake (*Micrurus fulvius*) and the Western Coral Snake (*Micruroides euryxanthus*), two of the world's most beautifully coloured serpents. These snakes, found in the hot southern States from Carolina to Mexico, have rings of brilliant red, yellow and ebony black circling their bodies. A few similarly coloured harmless snakes belonging to the king snakes group and living in similar areas, are often referred to as false coral snakes. These imitators can be readily identified as their undersides are not marked with rings of colour like the true coral snakes, but are plain cream or slightly blotched. There is also a different colour arrangement; the poisonous snakes have yellow and red colour bands adjoining, whereas the false corals have a black band separating the yellow and red.

The venom of the coral snakes is a neurotoxic (nerve poison) of an exceedingly potent type; it is probably mainly due to the fact that these snakes possess such small mouths and fangs that more accidents to humans do not occur from them. In the past, half the number of recorded bites from these snakes have had fatal results. They are mainly burrowing snakes, often hiding up in old rotten tree stumps or buried down in loose moist soil near woodlands, where they hunt for the skinks upon which they chiefly feed. In captivity they seem to prefer to eat small snakes, but they are 'difficult' reptiles to sustain in a vivarium.

Many of the harmless snakes are more adaptable and suitable for captivity than any of the venomous ones, and the king snakes and milk snakes of the genus *Lampropeltis* are particularly good examples. In common with most of the North American constrictors, these are useful rodent feeders, generally eating the mice and rats so destructive to food crops. They are also cannibalistic and often kill and devour some of the most dangerous snakes without harming themselves. They have been known to attack rattlesnakes and swallow specimens nearly as large as themselves. Their

Below: this Common Box Turtle (Terrapene carolina) *has withdrawn its legs into its shell.*
Right: American Alligator (Alligator mississipiensis) *about to eat a turtle.*

Below: the Arizona Coral Snake (Micruroides euryxanthus) *also lives in deserts of New Mexico and Mexico.*
Below right: the Common Water Snake (Natrix sipedon) *swims well and feeds on frogs and fish.*
Bottom left: Eastern King Snake (Lampropeltis getulus) *female guarding her eggs by curling around them.*
Bottom right: Ribbon Snake (Thamnophis sauritus) *a common aquatic snake of south-east Canada and the eastern United States.*
Far right: Water snakes live mainly in lakes and rivers. They are harmless to man. This is a Glossy Water Snake (Natrix rigida).

method of attack is to strike and thus obtain a hold with their jaws near the head of the other snake, and when this is accomplished, quickly to throw a few coils around until asphyxiation takes place.

All the snakes of this group are attractively patterned with spots, stripes and bands on their polished skin. The largest, such as the brown-striped Californian King Snake (*Lampropeltis californiae*) or the nearly black chain-marked Eastern King Snake (*Lampropeltis getulus*) can reach a length of seven feet, while some of the smaller species barely grow more than two feet. Towards humans they are mainly friendly, but in country districts the farming people are not always kindly disposed to the Milk Snakes (*Lampropeltis triangulum*). They are maligned as stealers of milk from the cows, a ridiculous accusation as it would be an impossible feat for the snakes to obtain milk from a cow, and in any case, they do not like drinking it, they prefer water. Yet they are still committed to the name of Milk Snake.

It is curious how common names have been acquired by many of the

snakes. Some are simply descriptive such as the group known as the garter and ribbon snakes (*Thamnophis spp.*). These are slim snakes, rarely growing to more than about thirty inches, and the dozen or more species are found from Canada to Central America. They are generally striped and bear a resemblance to the ornamental garters and ribbons worn years ago. Although not so aquatic as the water snakes (*Natrix spp.*) they are usually to be found near streams and pools, the damp areas where they can obtain the frogs, fish and salamanders upon which they feed. One of the commonest is the Red-sided Garter Snake (*Thamnophis sirtalis parietalis*) which is also one of the most attractive. It has brilliant scarlet sides and orange stripes along the whole length of its body.

The water snakes are larger and more robust and may be found in any stretch of water which affords them sufficient cover and food. It is surprising to find that some species are numerous, especially as the angling fraternity often wages war against them. It is not a good policy, since the water snakes eat the sick and ailing fish – which are easier to catch – before tackling the healthy specimens, and in this way the quality of the fishing is improved.

Mimicry is frequently met with in the reptiles, and the hog-nosed snakes (*Heterodon spp.*) are outstanding examples of harmless snakes imitating venomous ones. The hog-nosed snake dilates its neck and hisses in a manner reminiscent of the cobras of the Old World. It is also an expert at feigning death. If it finds escape from an adversary impracticable it will twist and turn with its mouth open as if in the

throes of convulsion, finally lying on its back in a death-like attitude. These are interesting snakes to keep as pets. When first caught they provide all the thrilling theatricals of venomous snakes, thrusting their heads forward as if to bite; but they never bite humans. After a very short time they become tame and appear to enjoy being carefully handled. Their diet consists of toads; I have known some individuals to eat frogs, but they cannot be fed entirely on frogs for long periods and remain fit. They evidently derive essentials from the warty skin and glands of the toads.

Two groups of snakes not yet mentioned are widely found all over the continent and vary a great deal in their different species and habitats.

The first group consists of the whip-snakes and racers. Generally long and slender, they are the fastest moving of all the American snakes. One or two species may grow to nearly eight feet in length and one, the Indigo or Corais Snake (*Drymarchon corais*) has been known to grow even longer in tropical America. It is a polished blue-black and makes an impressive sight as it glides along, often ignoring humans. It is not an aggressive snake, and although it will feed on almost anything that moves, including the large venomous snakes, it will soon become friendly. The Indigo Snake is not constricting as one would expect a snake of this type to be, but it relies on its powerful jaws to subdue its prey until swallowing is completed, and sometimes its food is still alive as it is engulfed. The racers and whip-snakes are often to be found in open prairie country, and this is where the Western Coachwhip (*Masticophis flagellum testaceus*) which feeds on the large green grasshoppers, is at home. The Eastern Coachwhip (*Masticophis flagellum flagellum*) which often shares the

Left: a young American Alligator (Alligator mississipiensis) *about 11 inches long. Alligators live in swamps in south-eastern United States.*
Right: Corn Snake (Elaphe guttata) *climbing up the trunk of a pine tree.*

burrows of the gopher tortoises in Carolina, is equally at home in and around the swamps of Florida.

The rat snakes (*Elaphe spp.*) make up the second large group. These are constrictors which largely feed on rodents though they are not adverse to eating birds, and some species eat birds' eggs. The rat snakes can be recognized by their flattened underside, which nearly forms a right angle with the sides of the body. Some are known as chicken snakes. The most beautiful is the Corn Snake (*Elaphe guttata guttata*), a four-foot snake which is often brilliant orange, with red blotches along its back having black pencilled edges. This snake and several other species of rat snake are partly arboreal

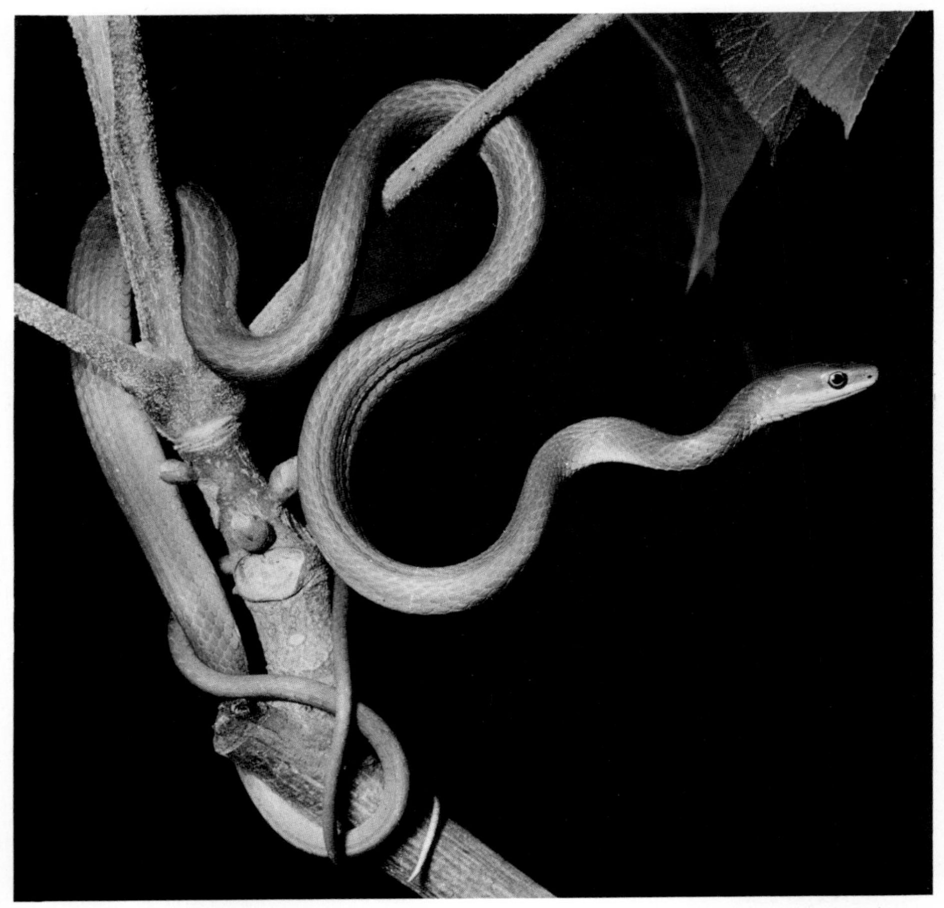

Left: Common Hog-nosed Snake (Heterodon platyrhinos) *of eastern North America.*
Below left: the Eastern Glass Snake (Ophisaurus ventralis) *is not a true snake but a legless lizard.*
Right: the Rough Green Snake (Ophoedrys aestivus) *often climbs bushes and low trees where it is hard to see.*
Below: Eastern Racer (Coluber constrictor) *about to swallow a young nestling Eastern Cardinal* (Richmondena cardinalis).

Left and below: the American Alligator (Alligator mississipiensis).
Right: Eastern Box Turtles (Terrapene carolina). *The lower shell (plastron) has a crosswise hinge, so that the shell can be completely closed.*

in their habits and individuals have been known to reside high up in the trunks of hollow trees, as do the American members of the *Boidae* family.

This family of giant snakes is represented in North America by two species of small snakes. The Rubber Snake (*Charina bottae*) and the Rosy Boa (*Lichanura roseofusca*) are both boas of two or three feet which live in the humid districts of western North America. The first has the appearance and feel of rubber. The second is similar but is a soft pinky colour. All the boas bear live young which have to fend for themselves and only a percentage reach maturity. Very large families with only a few growing to adulthood are common among the reptiles; this applies particularly to some of the chelonians which lay hundreds of eggs, and also to the alligators and crocodiles.

In southern parts of the continent both alligators and crocodiles occur. Up to the beginning of this century alligators were to be seen through many of the waterways and pools throughout the southern states, from North Carolina to Florida and across as far as the Rio Grande. But their hides began to be recognized as a valuable commodity and they were hunted and persecuted to such an extent that in many areas they were completely eradicated. The skins fetched high prices and even the small trading posts which were set up for the hunters at one time handled many thousands of skins each year. It is certain that these wonderful saurians would have gone the way of the Passenger Pigeon and become extinct if protection laws had not been passed. Nowadays, though not common reptiles, they are once more breeding in the many waters from which they had disappeared.

The American Alligator (*Alligator mississipiensis*) has been known to attain thirteen feet in length; but ten-foot males and six or seven-foot

females are now considered large. Very large specimens may be sixty years old, or even older. They are not aggressive towards humans, except at breeding time. If they have eggs in their huge nests – up to eight feet in diameter and built waist high among the reeds – then they may be belligerent. Otherwise they will lie on the river bank or submerge just out of sight. The males will fight among themselves during early summer, and often inflict serious and sometimes fatal injuries on each other. The number of eggs laid largely depends upon the size of the female, small ones laying up to twenty, but more mature specimens producing forty or more the size of a goose egg.

The American Crocodile (*Crocodylus acutus*) is found in the south of Florida where it overlaps with the alligator, but it requires more tropical conditions and is most numerous in Central America and the West Indies. It has a longer, more pointed snout, and can grow as long as twenty feet though few reach this size nowadays. Outside Florida the crocodile does not have the same protection as the alligator, and it can only be hoped that its numbers will not diminish further than they have already and that it will make its come-back like the alligator and some of the terrapins.

In America the chelonians are all grouped under the collective term of 'turtles'. For this chapter the method which is generally adopted in Britain is followed; only the marine creatures are referred to as turtles, and the terrestrial and fresh-water species as tortoises and terrapins respectively. In general, the tortoises have high domed shells and stumpy legs, the terrapins flatter shells, flatter legs and more developed feet, and the turtles large paddle-like legs, ideally suited to their marine existence.

There is a unique group of three species of land tortoises living in the Southern States and Mexico known as gopher tortoises. Many kinds of tortoises dig shallow holes in which to shelter or hibernate, but no others in the world are such tunnellers as these. The Eastern Gopher Tortoise (*Gopherus polyphemus*) often digs burrows thirty feet or more long and up to six feet in depth in which it spends its nights in company with snakes, frogs and mammals. They live in colonies and large patches of ground are undermined with their workings. The Desert Tortoise (*Gopherus agassi*) does not excavate to the same extent but often tunnels up to eight feet deep which gives it protection from both the very high and the low temperatures of the deserts where it lives, in parts of California, Arizona and Mexico.

The box tortoises are often kept as pets throughout North America. They may be encountered anywhere on the eastern side of the U.S.A. where there are woods or dampish open country.

They love to wallow during the hot weather in muddy marshes and will attempt to escape from an intruder by entering water and submerging. They are called box turtles from their hinged plastron, both ends of which can close to produce tight-fitting 'lids' to protect the head and limbs. They are small tortoises, the largest, *Terrapene major*, scarcely reaching seven inches in length. This species is rather dull plain brown, while others, notably the Ornate Tortoise (*Terrapene ornata*) has bright gold radiating lines and spots.

Far left: the Chicken Turtle (Deirochelys reticularia) *lives in still waters but frequently ventures on to land.*
Right: Chuckwallas (Sauromalus obesus) *feed on the flowers and fruits of cacti. They are about 16 inches long.*
Below: Beaded Lizard (Heloderma horridum) *one of the two known species of poisonous lizards. It inhabits the deserts of Arizona, Utah, Nevada and Mexico. The bite is not usually lethal to man.*

Top: female Ground or Banded Gecko (Coleonyx variegatus) *shedding her skin.*
Above: the young Five-lined Skink (Eumeces fasciatus) *is dark brown with stripes but the adults become lighter and the stripes fade.*
Right: the Texas Horned Lizard (Phrynosoma cornutum) *is also called the Horned Toad.*

The greater number of terrapins are harmless and docile animals when confronted. Some of them just withdraw their head into their shell with a soft hiss; others will try to scamper away or dive into a pool to escape, as the sliders (*Pseudemys spp.*) do; just a few may give a nip if handled and the musk terrapins (*Kinosternum spp.*) exude a most foul odour from glands near their tail. There is one sub-family of terrapins however, the Chelydrinae, snapping terrapins, which can be described as being bad tempered, and when adult these can inflict a serious bite. The Common Snapper (*Chelydra serpentina*) may be found in any large stretch of open water from Southern Canada to Central America. Snappers cannot be mistaken, except perhaps when very young, for any other terrapin. They have a large head with a hook-like projection on the upper jaw giving them the appearance of a bird of prey. The horny, rough brown shell has three ridges down its length and its strong thick tail has a saw-like projection along the top. The Alligator Snapper (*Macroclemys temmincki*) has a slightly different shell and tail

formation and a more restricted range in the south-east of the United States. It is the world's largest terrapin and has been known to reach over two hundred pounds in weight.

The most beautiful of the 'hard-backs' are the Painted Terrapins (*Chrysemys picta*) and some of the cooters or sliders (*Pseudemys spp.*). Some are striped with yellows and greens, others have red spots and splashes and lacy patterns on the shell and body. A few have coral red plastrons or bright red edging. There is an interesting and colourful courtship display by some species in the spring. Just previously, the male grows very long claws on the front feet and these he waves and weaves about as he swims, facing the front of his intended bride like a fan dancer. If the response from the female is satisfactory and pairing takes place, fertile eggs may be laid each season for several consecutive years from a single mating.

The soft-shelled terrapins are entirely aquatic and all but one species are river dwellers, found in waterways from the St. Lawrence to Florida. The exception is the Southern Softshell (*Trionyx ferox*) which lives in ponds and drainage ditches in Georgia and Florida. Most of their time is spent in the mud and soft sand just below shallow water. Until they move, they can hardly be recognized as animals, and appear to be slabs of gritty mud. Sometimes they will climb up the bank to lie and sunbathe, spreading out their flat, heavily clawed feet and long concertina-like head and neck. They can give a very unpleasant bite, but are normally nervous of humans and will hurriedly slide back into the water on their approach.

The marine turtles are the largest of the chelonians found around the North American continent. Their normal home is in the warm tropical seas but numbers follow the Gulf Stream northwards and every summer a few are seen drifting off Cape Cod. But they do not nest so far north. The Loggerhead (*Caretta caretta*) appears to be the hardiest, and this large brown species which may weigh 800 pounds, sometimes lays its eggs on secluded beaches of Virginia, but for the other four species found round these shores, Florida is generally the northern limit.

The largest of the turtles is the huge Leatherback (*Dermochelys coriacea*). It is a turtle of the open sea and has been reported to reach nearly a ton. It has been known to nest in the West Indies and the Florida coast, but it is rare in these parts.

What intriguing names there are among the lizards of North America. The racerunners or whiptails and the swifts are all agile lizards which live up to their name. The Chuckwalla (*Sauromalus obesus*) sounds an appropriate name for this rather obese and slow lizard which only feeds on plants and flowers. Actually, the word means a reptile in one of the Indian dialects. The lizards are the most adaptable of the reptiles, and show more variety to suit their way of life and habitat than any other group. Few species in North America are found in damp marshy parts; the majority live in the drier areas, in arid land where little else can survive.

In this type of terrain are found the world's only two venomous lizards, the Gila Monster (*Heloderma suspectum*) and the Mexican Beaded Lizard (*Heloderma horridum*). Unlike the venomous snakes which have their poison glands located in the upper jaws, these lizards have theirs in the lower jaw and inject the venom by chewing with their powerful jaws. They do not deserve the evil reputation they have, as they are sluggish animals and are rarely seen near townships. Bites have been inflicted on humans, but there has been only one recorded death from their bite in fifty years. Adult specimens may grow to a little more than two feet in length. They have heavy bodies, short powerful legs equipped with strong claws, and a thick fleshy tail which is a fat reserve during times of scarcity. The Gila Monster is salmon pink with black lacy markings, the Beaded Lizard is nearly black with yellow lacework, and their skin has the appearance of beadwork on the handbags of the Victorian era. They are rare reptiles and are only known in the deserts of Nevada, Utah, Arizona and Mexico.

Above: Fowler's Toad (Bufo woodhousei fowleri) *'singing' to attract a female. Below: the Green Tree-frog* (Hyla cinerea) *is bright emerald green in warm weather but on cool days it is slate grey. Right: Green Frog* (Rana melanota).

There are many other interesting lizards in the south-western deserts, but none are more grotesque than the horned toads (*Phrynosoma spp.*) of which there are about fifteen species. These are flat sand-coloured lizards with bodies half as wide as they are long, covered with rows of small soft spines. The head is crowned with large horn-like scales giving the creature the appearance of a miniature six-inch oriental dragon. They never bite and are completely harmless but if they are handled they often feign death, and they also have the unique ability of ejecting, to several feet away, very fine jets of blood from the corners of the eyes. Among the other iguanid lizards the most outstanding is the Desert Iguana or Crested Lizard (*Dipsosaurus dorsalis*) which lives in open deserts of the south-western United States.

The American 'Chameleon' or Anole (*Anolis carolinensis*) is not a true chameleon but is able to change colour just as quickly from green to browns and greys. The male of the species has a dewlap under the chin like a small orange fan, and with this he courts the female by bobbing his head up and down as he sits on a branch in the sun. To enable the Anole to climb on the smooth trunks of trees it has small pads on its feet similar to those of the geckos, which also have pads of small ridges on the ends of each toe. These are very effective in gripping, even on glass-like surfaces, and are essential to the lizards if they are to catch settling mosquitos.

The burrowing skinks do not need help in climbing, as much of their life is spent underground, but they have shovel-like snouts to burrow with, and many of the lizards have segmented tails which easily break off when gripped by a predator, so nature has given the lizards many aids to survive. It is to be hoped that all the reptiles will manage to do so; so much can be learnt from these fascinating creatures.

INDEX

References to illustrations in italic

Photographic Acknowledgments Rod Allin 68 B–69; Heather Angel 18; Ronald Austing 37 BL, 38 R–39, 44 T, 44 B, 48 L, 48 TR, 51, 52 C, 54 T–55 TL, 54 BL, 54 BR; Des and Jen Bartlett 65 T, 70 T, 72 T, 104 T, 104 B, 108; Bureau of Sport Fisheries and Wildlife, Washington DC, 22 CR, 28 BR–29, 31 T, 34–35, 82 B, 92 TL, 106 B–107; Jane Burton 45 B, 65 B, 105 TL; Robert Bustard 119 T, 120 T; Lynwood Chase 83 T; W. T. Davidson 10 C, 11 T, 17 C, 109; Evan J. Davis 45 T; Ken Denham 10 B; J. Dermid front endpapers, 40 TL, 41 R, 55 BR, 58, 77 TR, 98–99, 100 TL, 101 C, 110 TR, 111, 112, 113, 114 B, 115 T, 115 B, 116 T, 117 T, 118, 120 CL, 120 CR–121, 122 T, 122 B, 105 B; Jerry Focht 55 C, 110 BR; Wallace Grange 31 B; James Hancock 25; R. B. Hoit 8–9; D. C. Honsowetz 84; Eric Hosking 22 L, 27 T, 28 T, 28 BL, 52 TR; William Howes 19; H. Roy Ivor 52 CL; Karl W. Kenyon 90 B; Russ Kinne 11 B, 12, 13, 15 T, 15 BR, 16–17 L, 17 T, 47, 48 BR, 50 B, 60, 61, 86 B, 91 BL, 123; G. E. Kirkpatrick 24 CR, 42–43 L, 43 T, 43 B, 70 B, 93 B, 103, 119 B; Frank Lane 82 T, 85; John Markham 50 T, 100 TR–101 T, 100 BL–101 BL, 101 BR; Marineland of Florida 14, 91 BR; Tom McHugh 36; Wilford Miller 73, 74–75, 77 TL, 80, 81 T, 81 B, 88–89 L; A. L. Mordon 24 L; Chas. J. Ott 32, 37 R, 68 T, 72 B, 77 B, 97, back endpapers; E. D. Park 92 BL; Roger Tory Peterson 40 BL, 40 R–41 L; Photo Researchers 114 T; Rex Features 62 T; Dick Robinson 76; Leonard Lee Rue 22 BR, 24 TR, 24 BR, 30, 46, 49, 52 L, 52 BR–53, 55 C, 59, 62 B, 63, 64 T, 64 B, 66 TL, 66 BL, 66 R, 67 L, 67 R, 71, 83 B, 87 T, 87 B, 89 T, 93 T, 95 CR, 95 BR, 96 R, 102 T, 102 B, 106 T, 110 BL, 116 B–117 B; Hugo H. Schroder 22 TR, 23, 26–27 L; Barrie Thomas 38 L; Grace A. Thompson 105 TR; Tierbilder Okapia 78 T, 79 B, 86 T, 89 B, 92 TR, 95 BL; Tiofoto 90 T–91 TL, 96 L; Douglas P. Wilson 10 T, 15 BL; Joe Van Wormer 20–21, 27 BR, 37 T, 56–57, 66 BR, 79, 94–95 TL; Z. F. A. 33, 100 BL.
Map drawn by Denys Ovenden.

Front endpapers: five young Opossums (Didelphis marsupialis) *about 90 days old, on a blue beech tree.*

Back endpapers: Barren Ground Caribou (Rangifer tarandus groenlandicus) *cow and calf crossing a glacial stream in Alaska.*